Improving the Social Security Representative Payee Program

Serving Beneficiaries and Minimizing Misuse

Committee on Social Security Representative Payees

Division of Behavioral and Social Sciences and Education

NATIONAL RESEARCH COUNCIL
OF THE NATIONAL ACADEMIES

THE NATIONAL ACADEMIES PRESS
Washington, D.C.
www.nap.edu

THE NATIONAL ACADEMIES PRESS 500 Fifth Street, N.W. Washington, DC 20001

NOTICE: The project that is the subject of this report was approved by the Governing Board of the National Research Council, whose members are drawn from the councils of the National Academy of Sciences, the National Academy of Engineering, and the Institute of Medicine. The members of the committee responsible for the report were chosen for their special competences and with regard for appropriate balance.

This project was supported by Award No. SS00-04-60082 between the National Academy of Sciences and the Social Security Administration. Any opinions, findings, conclusions, or recommendations expressed in this publication are those of the author(s) and do not necessarily reflect the views of the sponsors.

International Standard Book Number-13: 978-0-309-11100-3
International Standard Book Number-10: 0-309-11100-5

Additional copies of this report are available from the National Academies Press, 500 Fifth Street, N.W., Lockbox 285, Washington, DC 20055; (800) 624-6242 or (202) 334-3313 (in the Washington metropolitan area); Internet http://www.nap. edu.

Printed in the United States of America.

Suggested citations: National Research Council. (2007). *Improving the Social Security Representative Payee Program: Serving Beneficiaries and Minimizing Misuse.* Committee on Social Security Representative Payees, Division of Behavioral and Social Sciences and Education. Washington, DC: The National Academies Press.

THE NATIONAL ACADEMIES
Advisers to the Nation on Science, Engineering, and Medicine

The **National Academy of Sciences** is a private, nonprofit, self-perpetuating society of distinguished scholars engaged in scientific and engineering research, dedicated to the furtherance of science and technology and to their use for the general welfare. Upon the authority of the charter granted to it by the Congress in 1863, the Academy has a mandate that requires it to advise the federal government on scientific and technical matters. Dr. Ralph J. Cicerone is president of the National Academy of Sciences.

The **National Academy of Engineering** was established in 1964, under the charter of the National Academy of Sciences, as a parallel organization of outstanding engineers. It is autonomous in its administration and in the selection of its members, sharing with the National Academy of Sciences the responsibility for advising the federal government. The National Academy of Engineering also sponsors engineering programs aimed at meeting national needs, encourages education and research, and recognizes the superior achievements of engineers. Dr. Charles M. Vest is president of the National Academy of Engineering.

The **Institute of Medicine** was established in 1970 by the National Academy of Sciences to secure the services of eminent members of appropriate professions in the examination of policy matters pertaining to the health of the public. The Institute acts under the responsibility given to the National Academy of Sciences by its congressional charter to be an adviser to the federal government and, upon its own initiative, to identify issues of medical care, research, and education. Dr. Harvey V. Fineberg is president of the Institute of Medicine.

The **National Research Council** was organized by the National Academy of Sciences in 1916 to associate the broad community of science and technology with the Academy's purposes of furthering knowledge and advising the federal government. Functioning in accordance with general policies determined by the Academy, the Council has become the principal operating agency of both the National Academy of Sciences and the National Academy of Engineering in providing services to the government, the public, and the scientific and engineering communities. The Council is administered jointly by both Academies and the Institute of Medicine. Dr. Ralph J. Cicerone and Dr. Charles M. Vest are chair and vice chair, respectively, of the National Research Council.

www.national-academies.org

Dedication

This report is dedicated to the memory of committee member Eileen P. Sweeney, who died June 13, 2006.

Eileen was a nationally recognized expert on issues affecting people with disabilities who receive federal welfare benefits. From the time she was in college through her work as a senior fellow at the Center on Budget and Policy Priorities in Washington, DC, she sought to improve the lives of children, battered women, senior citizens, poor people, and people with disabilities. In 2005 and early 2006, she cochaired the Social Security Task Force of the Consortium for Citizens with Disabilities, a coalition of national organizations advocating on behalf of the 54 million Americans with disabilities.

This report and the committee's recommendations were deeply influenced by Eileen's work, values, and sensitivity and commitment to the rights of all Social Security beneficiaries. The committee was fortunate to have Eileen as a member.

Acknowledgments

Many people and groups made significant contributions to this report. Without their willing help, expertise, and dedication to completing the project's many tasks in a quality and timely manner, we would have a far less complete report.

Our first thanks go to the staff of the U.S. Social Security Administration (SSA), who helped us understand the Representative Payee Program and patiently answered our many questions about policies and regulations. The staff was also invaluable in helping us secure clearance from the U.S. Office of Management and Budget (OMB) for the committee's survey and in diligently and expediently helping us with obtaining crucial security clearances for Westat employees, who conducted the survey, including more than 100 field interviewers. Security clearances were required because of the confidentiality of information that was collected in the survey of representative payees and beneficiaries.

SSA staff graciously set up work space for us at their Rolling Road Commerce Center in Baltimore. Here, they gave us access to the Representative Payee System (RPS). They tutored us on the RPS and facilitated file acquisitions and specific administrative data requests. They also made arrangements for us to visit field offices, regional offices, and processing centers.

Among the SSA staff, many individuals deserve our specific thanks. Michael Zambonato was not only the SSA project officer, but he readily gave us his expertise and insights in many situations.

Sherrye Walker, director of the Office of Beneficiary Determinations, and Patti Gavin on her staff made sure we had everything we needed to do our work. Marg Handel, supervisor of the Representative Claimant Team in the Office of Payment Policy, assisted us with extraction of monthly benefit amounts for our study participants.

For their help with the RPS, the case study of misusers, and the lump-sum payments analysis, we thank Kevin McCahill (team leader, now retired), Barbara Benjamin, Lynn Brown, Chris Garcia, Pat Gregus, Bryan Mueller, and Paul Sapia on the Representative Payment Monitoring and Evaluation Team, and on the Representative Payment Selection and Systems Team, Steve Auerback (team leader), Lavern Alston, Kevin Brennan, and Ray Hairston. We especially thank Betsy Byrd for not only her help on these projects, but also her assist with numerous special requests.

Debra Acord on the Special Projects Team and Mary Gibson in the Office of Eligibility and Enumeration Policy and the Eligibility and Evidence Team cheerfully came through for us again and again. Without their ingenuity and resourcefulness, we could not have successfully carried out the case study of misusers or made our invaluable site visits.

During the initial phases of our study, Carol Musil (now retired) was instrumental in easing our way through many SSA protocols and "how to get stuff done."

Liz Davidson and Faye Lipsky on the Reports Clearance Team were instrumental in obtaining the timely OMB clearance for the survey.

We also thank Thomas Bell of Social and Scientific Systems, Inc., who worked under contract with the National Academies to perform onsite computer work at SSA. He worked in manipulating large administrative databases, retrieving much needed information from the SSA mainframe system with only minimal specifications, and creating sample files. He was of critical continuous support to our analysis efforts.

Westat conducted the survey for this report under contract with the National Academies. We are grateful to the many people who contributed to this very successful phase of our study, including co-project leaders Scott Crosse and Diane Cadell. We also thank staff from various areas at Westat, including Richard Sigman, Wendy Kissin, Michelle Scheele, Vicky Klementowicz, Mike Rhoades, Sheryl Wood, Joe Gertig, Tory Castleman, Miriam Aiken, and Ellen Herbold.

Finally, enormous thanks are due to the committee members and staff, who truly represent the best of the National Academies. The members brought extraordinary knowledge, experience, and commitment to the task, and they gave unstintingly of their time. With almost a dozen face-to-face meetings and dozens of conference telephone calls, they contributed enthusiastically and critically to every aspect of the project, from the overall design of the tasks to the final conclusions and recommendations.

Keeping everything organized and on time was our invaluable study director, Bud Pautler. At every stage of the project, he laid out the plans and the questions, arranged innumerable site visits, drafted text, and did myriad other tasks that made our work possible and fun. He was assisted throughout by Kirsten West, on an interagency personnel assignment to the National Academies from the U.S. Census Bureau. Her quiet competence was a constant support for us. No question or task was too obscure for her; if some information or calculation was requested and was possible, she found it or did it. He was also assisted by Linda DePugh who made the meeting logistics work, processed committee travel arrangements, and helped with the production of this report. I am also indebted to Genie Grohman, who made significant editorial suggestions. I am deeply grateful for the opportunity to have worked with this group of talented and committed individuals.

This report has been reviewed in draft form by individuals chosen for their diverse perspectives and technical expertise, in accordance with procedures approved by the National Research Council's (NRC) Report Review Committee. The purpose of this independent review is to provide candid and critical comments that will assist the institution in making its published report as sound as possible and to ensure that the report meets institutional standards for objectivity, evidence, and responsiveness to the study charge. The review comments and draft manuscript remain confidential to protect the integrity of the deliberative process. We thank the following individuals for their review of this report: Lu Ann Aday, Public Health and Medicine, The University of Texas, Houston; Peter Blanck, Burton Blatt Institute, Syracuse University; Eric Elbogin, Department of Psychiatry and Behavioral Sciences, Duke University; Edward Lawlor, George Warren Brown School of Social Work, Washington University in St. Louis; Robert A. Moffitt, Department of Economics, Johns Hopkins University; John Monahan, School of Law, University of Virginia; David Weir, Institute for Social Research, University of Michigan.

Although the reviewers listed above have provided many constructive comments and suggestions, they were not asked to endorse the conclusions or recommendations, nor did they see the final draft of the report before its release. The review of this report was overseen by Tim Smeeding, Center for Policy Research, Maxwell School, Syracuse University, and Joe Newhouse, John D. MacArthur Professor of Health Policy and Management, Harvard University. Appointed by the NRC, they were responsible for making certain that an independent examination of this report was carried out in accordance with institutional procedures and that all review comments were carefully considered. Responsibility for the final content of this report rests entirely with the authoring committee and the institution.

Barbara A. Bailar, *Chair*
Committee on Social Security
Representative Payees

Contents

Executive Summary

More than 7 million recipients of Social Security benefits have a representative payee—a person or an organization—to receive or manage their benefits. These payees manage Old Age, Survivors and Disability Insurance funds for retirees, surviving spouses, children, and the disabled, and they manage Supplemental Security Income payments to disabled, blind, or elderly people with limited income and resources. More than half of the beneficiaries with a representative payee are minor children; the rest are adults, often elderly, whose mental or physical incapacity prevents them from acting on their own behalf, and people who have been deemed incapable under state guardianship laws. The funds are managed through the Representative Payee Program of the Social Security Administration (SSA). The funds total almost $4 Billion a month, and there are more than 5.3 million representative payees.

In 2004 Congress required the commissioner of the SSA to conduct a one-time survey to determine how payments to individual and organizational representative payees are being managed and used on behalf of the beneficiaries.[1] To carry out this work, the SSA requested a study by the National Academies, which appointed the Committee on Social Security Representative Payees. This report is the result of that study.

The commissioner set four objectives for the study: (1) assess the

---

[1]The Social Security Protection Act (SSPA) of 2004 (P.L. 108-203).

extent to which representative payees are not performing their duties in accordance with SSA standards for representative payee conduct, (2) learn whether the representative payment policies are practical and appropriate, (3) identify the types of representative payees that have the highest risk of misuse[2] of benefits, and (4) find ways to reduce the risk of misuse of benefits and ways to better protect beneficiaries.

A national survey was conducted for the committee by Westat, Inc. It involved interviews with more than 5,000 representative payees across the country; for half that group, about 2,500, the beneficiaries were also interviewed. As mandated by Congress, the committee restricted the national survey to individual payees serving fewer than 15 beneficiaries and non-fee-for-service organizational payees serving fewer than 50 beneficiaries. The sample included payees and beneficiaries from the 48 contiguous states for representative payees managing at least $50. With these restrictions, the study population was approximately 3.5 million payees.

In addition to the national survey, the committee visited SSA field offices, conducted an in-depth study of known misusers, examined lump-sum payments,[3] and examined violations of SSA representative payee rules that do not constitute misuse. The committee also developed, with a forensic auditor, a reinterview process designed to detect misuse of a beneficiary's benefits. The sample of payees reinterviewed was selected based on the scoring of hypothesized characteristics of misusers.

PAYEE PERFORMANCE

The committee found that the majority of representative payees perform their duties well and generally understand their responsibilities. Payees and their beneficiaries appear to maintain reasonably frequent contact and discuss basic needs: about 86 percent of payees report being in touch with their beneficiary at least once a week. Both payees and beneficiaries report high levels of satisfaction with payee performance: almost 95 percent of beneficiaries report they are "satisfied" or "very satisfied" with their payees. Payees and beneficiaries have high levels of understanding—more than 90 percent of both groups—of the basic duties and responsibilities of representative payees.

However, many payees are unaware that they should place unused funds in a savings account. Because Social Security benefits are often the

---

[2]"Misuse occurs in any case in which the representative payee receives payment under this title for the use and benefit of another person and converts such payment, or any part thereof, to a use other than for the use and benefit of such other person." SSPA, § 1631(a)(2)(A)(iv).

[3]Most lump sums are retroactive benefits, i.e., benefits accrued prior to an award. Lump sums might also accrue while payment is suspended for selection of a new payee.

only source of income for most beneficiaries, it is important that payees use funds for the needs of beneficiaries and conserve them whenever possible. More broadly, the program does not require careful accounting and reporting by payees, nor does the current system appear to be useful in detecting possible misuse of benefits by payees.

CONCLUSION As a practical matter, the Social Security Administration does not require representative payees to carefully account for dollars spent in each category on the annual accounting form as long as the total amount spent is approximately the same or greater than the total amount of benefits received.

CONCLUSION The Social Security Administration does not apply special monitoring to payees of beneficiaries who receive lump-sum payments.

CONCLUSION Large lump-sum payments often appear to be spent on items or persons that would not be approved by the Social Security Administration and would be considered a misuse of funds.

RECOMMENDATION 3.1[4] The Social Security Administration should strengthen its efforts to encourage representative payees to save money for beneficiaries and to enforce the requirement that the saved money is put in a specified savings account.

RECOMMENDATION 4.1 The Social Security Administration should give special scrutiny to representative payees who receive lump-sum payments.

In the committee's in-depth study of payees who had been identified by SSA as misusers, it learned that a small number were still serving as payees, a practice that should be discouraged whenever possible.

RECOMMENDATION 4.2 The Social Security Administration should develop new procedures and policies that prevent the routine reappointment of a representative payee who has been documented as a misuser or a continued violator of Social Security Administration policies and rules.

---

[4]The recommendation number refers to the chapter in which the recommendation appears.

PREVENTION AND DETECTION OF MISUSE

The SSA's main tool to discover misuse is the annual accounting form; SSA also relies on beneficiaries, relatives, and other concerned persons to bring misuse to its attention. In addition, the Office of the Inspector General uses small, simple random samples of representative payees to detect misuse. The annual accounting form does not request sufficient information for detecting misuse and is not effectively reviewed or stored in SSA systems. In the cases of the personal reports, the cases of suspected misuse are often not investigated or documented as misuse and generally result in a change in payee with the notation, "more suitable payee found." The result of the audits by the Office of the Inspector General is that no cases of misuse have been found: SSA has officially reported the amount of misuse in the program to be less than 0.01 percent.

The committee developed a new approach of identifying potential misusers and interviewing them with a two-person team that included an auditor and a social scientist with the goal of improving the ability to detect misuse in samples. In an in-depth study of 76 cases selected using this new methodology, the committee found 16 (21 percent) misusers and 17 (22 percent) cases of possible misuse but for which there was insufficient information to confirm misuse. Applying the committee's methodology to the types of payees that the committee studied, more than 40,000 representative payees have many of the characteristics associated with misuse and warrant investigation. Among those estimated 40,000 payees, an investigation would probably find about 7,000 misusers and another 7,000 uncertain or potential misusers. The total number is still a small percentage of misusers in the population (about 0.2 percent), but it is significantly higher than the SSA estimate.

CONCLUSION Relying on beneficiaries or third parties to report misuse to the Social Security Administration is not a reliable or efficient primary strategy for detecting misuse.

CONCLUSION The methodology used by the Social Security Administration Inspector General does not detect misuse.

CONCLUSION The Social Security Administration does not discover misuse by using the annual accounting form.

CONCLUSION The characteristics identified in the committee's in-depth study of misuse are effective in targeting representative payees for auditing for the purpose of detecting misuse.

CONCLUSION The use of a specialized team of auditors was effective in uncovering misuse of funds by representative payees.

RECOMMENDATION 5.1 The Social Security Administration (including its Inspector General) should use probability sampling with targeted sample selection, using criteria associated with misuse of funds such as those suggested in this report, to audit representative payees who are more likely to be misusers.

RECOMMENDATION 5.2 The Social Security Administration should develop criteria associated with misuse of funds, such as those suggested in this report, to select and monitor representative payees.

RECOMMENDATION 5.3 The Social Security Administration should establish a team of experts, such as the audit teams used in the committee's study, to audit those payees who are suspected of misuse or who have been included in a targeted sample of potential misusers.

RECOMMENDATION 5.4 The Social Security Administration should redesign the annual accounting form to obtain meaningful accounting data and payee characteristics that would facilitate evaluation of risk factors and payee performance.

A preliminary attempt at redesign of the annual accounting form by the committee (but without cognitive testing) is presented in Appendix F.

The committee identified additional strategies to improve tracking of payee performance, including the use of debit cards and a more comprehensive support system to enable broader evaluation of payee behaviors.

RECOMMENDATION 5.5 The Social Security Administration should conduct a test of bank-account-linked debit cards for representative payees.

RECOMMENDATION 5.6 The Social Security Administration should initiate a research, development, and support function for the Representative Payee Program to promote quality and cost-effectiveness in its operations.

PROGRAM POLICIES AND PRACTICES

The committee found that the Representative Payee Program is generally meeting the needs of payees and beneficiaries. More than 96 percent of beneficiaries and payees agreed that the beneficiary needed help in managing their funds, and more than 97 percent of payees expressed a willingness to serve their beneficiary.

However, in assessing the procedures for selecting, training, supporting, monitoring, and terminating payees, the committee found several deficiencies. The implementation of existing field office procedures for those activities is constrained by staff resources and the lack of incentive systems for staff to spend time on important tasks. Moreover, the agency is not obtaining the information it needs to manage the program efficiently, to provide the best service possible to beneficiaries, and to detect cases of misuse.

Selection and Oversight of Payees

The committee found that there is a good deal of office-to-office variation in the payee selection process. This suggests a need to standardize the selection criteria and process. In addition, the committee found weaknesses in the program with respect to finding payees for at-risk beneficiaries such as people with mental illness, alcohol or substance abuse problems, severe disabilities and those who are homeless.

CONCLUSION It is difficult to find appropriate payees for at-risk beneficiaries. Fee-for-service payees may be better for at-risk beneficiaries because they are professionals and may be licensed and are better equipped to deal with situations posed by at-risk beneficiaries.

RECOMMENDATION 6.1 To help mitigate shortages of payees, the Social Security Administration should create a program to identify, train, certify, and maintain a pool of voluntary, temporary payees that are available on an as-needed basis. If such a program is authorized, the Social Security Administration should work with and obtain help from the courts and volunteer organizations in designing it.

RECOMMENDATION 6.2 Congress should authorize the Social Security Administration to expand the fee-for-service part of the program to include appropriate small organizations and individuals who are willing to serve as payees for at-risk beneficiaries: people with mental illness, alcohol or substance abuse problems, severe disabilities, and those who are homeless.

In general, the committee found that the process for determining if an individual is suitable to serve as a payee needs improvement. While nearly all payees serve willingly, a critical dimension of performance is ability. A reasonably large proportion of selected payees appear to be in economically unstable circumstances, have a criminal background, or have a prior substance abuse problem. Selecting such individuals as payees likely increases the risk of misuse.

CONCLUSION The Social Security Administration appoints some payees with characteristics that raise questions about their suitability as payees.

RECOMMENDATION 6.3 The Social Security Administration should screen potential payees (including organizational payees) for suitability on the basis of specified factors associated with misuse, particularly credit history and criminal background.

RECOMMENDATION 6.4 The payees of at-risk beneficiaries should be monitored more frequently and intensively than current protocols provide.

Payees who serve for multiple beneficiaries and who also operate room-and-board facilities or similar facilities have potential conflicts of interests and need to be monitored more closely. It is possible for a payee to operate several group homes with up to 14 beneficiaries in each and thus be treated as an individual payee by SSA. Moreover, state and local authorities may be unaware that the payee is operating several homes.

CONCLUSION The current designation of "individual payee" is too broad a category. The designation mixes payees who serve a single or even a few beneficiaries with payees who operate group homes for up to 14 beneficiaries. Individual payees who are owners or administrators of group homes have an inherent conflict of interest. Payees of this type require special monitoring.

RECOMMENDATION 6.5 The Social Security Administration should develop policies that define and treat as an organizational payee an individual who serves multiple, unrelated beneficiaries and who is also the owner, administrator, or provider of a room-and-board facility.

RECOMMENDATION 6.6 The Social Security Administration should reevaluate its policies that permit creditors and administrators of facilities to serve as payees.

Guardianship and the Fee-for-Service Program

Many aspects of the program involving guardianships and fee-for-service providers warrant further scrutiny from SSA because of conflicts among federal law, SSA policies, and state practices. There are also potential conflicts when beneficiaries have SSA-appointed payees who are different from the people who hold powers of attorney for them or who serve as their legal guardians or conservators.

CONCLUSION The guardianship and fee-for-service aspects of the program conflict with the congressional intent that individual payees not receive fees from Social Security funds. Although the Social Security Administration Program Operating Manual System provides policy guidance for allowing fees when there is court oversight, this broad allowance of such a practice is not in the best interests of beneficiaries and conflicts with legislative intent.

CONCLUSION Some beneficiaries have Social Security Administration-appointed payees who are different from the people who hold their power of attorney or serve as legal guardian, or conservator. This causes potential conflicts, violations of Social Security Administration rules, inefficiencies and inaccuracy in reporting, delays in payee selection, and duplication of effort.

CONCLUSION There is a lack of communication between the Social Security Administration and state courts with regard to beneficiaries who might have both a guardian and a representative payee. This lack of communication has led to misunderstandings as to the authority, or lack thereof, for paying fees for representative payee services.

RECOMMENDATION 6.7 The Social Security Administration should change the Program Operating Manual System to state that when a beneficiary already has an individual with power of attorney, a legal guardian, or conservator, there is a preference (with flexibility) for selecting that individual as the beneficiary's representative payee.

RECOMMENDATION 6.8 The Social Security Administration, in consultation with the states, should eliminate inconsistencies between state and federal practices regarding the calculation of payee fees and financial oversight.

Payee Training

The committee found that, after selection, the SSA does little to help payees perform the required functions and best serve their beneficiaries and the program. The payees must understand their duties and responsibilities, including details such as how to keep records, how to deposit benefits into separate accounts, and how to save money. Specific materials that would facilitate a detailed understanding of responsibilities, such as how to record expenses and income to monitor payments and their use, are not available to payees.

RECOMMENDATION 6.9 The Social Security Administration should provide comprehensive and formal training for representative payees.

RECOMMENDATION 6.10 The Social Security Administration should provide payees access to various types of well-advertised support in their activities. Such support could include: (1) dedicated field staff who can serve as contact persons for payees; (2) toll-free telephone numbers specifically for use by payees to seek assistance from SSA; (3) easily comprehensible brochures containing examples and explanations; (4) enhanced, easy-to-use FAQs and online learning tools; (5) guidance on how to meet accounting and document retention requirements; and (6) online guidance for payees to complete the annual accounting form.

Monitoring Payee Performance

The accounting process and the annual accounting form are the primary SSA strategies for monitoring payee performance. Although there is a perception that the process is a psychological deterrent to misuse, the process does not actually lead to the identification of misuse and the annual accounting form itself does not yield data for statistical evaluations of program performance.

CONCLUSION The statutory provision [42 U.S.C. § 405(j)(3)(A) (OASDI); 42 U.S.C. § 1383(a)(2)(C) (SSI)] that requires the commissioner of the Social Security Administration to establish and implement statistically valid procedures for reviewing the annual accounting forms creates a concomitant obligation to provide information for understanding and monitoring the performance of representative payees. This obligation is not being fulfilled.

CONCLUSION The filing of annual accounting forms (or the failure to file them) is not reconciled with any other administrative record so that a failure to file would bar a payee from continuing to serve in such a capacity.

CONCLUSION It is too easy for representative payees to learn that if they just fill out the accounting form with some plausible, but possibly inaccurate information, they will have complied with the program's reporting requirement and that there will be no follow-up or other consequences. Essentially, the current monitoring process is an "empty threat" that can easily be subverted and is an expensive administrative tool that does not yield the sort of data that are necessary to uncover misuse.

CONCLUSION The Social Security Administration does not have a method for systematically evaluating and validating the material it receives on the annual accounting forms. The data on the accounting form are not retrievable for statistical analysis and therefore, empirically based policies and regulations cannot be formulated. In addition, the Social Security Administration's legislative obligation to statistically tabulate the annual accounting form remains unfulfilled.

RECOMMENDATION 6.11 The Social Security Administration should reengineer the annual accounting form to ensure the usefulness of the data and their transferability into the Representative Payee System and other Social Security Administration information systems.

RECOMMENDATION 6.12 The Social Security Administration should store data from the annual accounting forms in an electronic database suitable for analysis.

RECOMMENDATION 6.13 The Social Security Administration should provide the option for payees to complete the annual accounting form online.

Another SSA strategy for monitoring payee performance is to rely on input from beneficiaries and third parties. However, even when such input is received, there are few formal rules or guidelines for staff to handle situations of alleged misuse. The committee found that the Representative Payee System (RPS) in its current form is not a useful mechanism for helping staff carry out their functions or for providing overall data on the program. The committee understands that plans are under way at SSA for revisions to the RPS; thus, this is an opportune time to consider staff and program needs for the RPS.

CONCLUSION Factors such as lack of incentives for staff to investigate misuse, perceived vagueness in the definition of misuse, and the complexity of interpersonal relationships between beneficiaries and their payees often lead claims representatives to find a more suitable payee rather than to formally determine misuse.

CONCLUSION Frequently changing custodial arrangements for beneficiaries who are children involve complicated situations that may facilitate payee misuse.

CONCLUSION The Representative Payee System is a badly flawed tool for case-by-case field use to evaluate prospective representative payees and to investigate problems with payees. Office-to-office autonomy regarding procedures for making entries into the Representative Payee System and a cumbersome and inefficient interface create an environment that encourages inconsistencies in the amount and quality of information available in the database. In addition, data quality concerns and incompleteness compromise the potential for the Representative Payee System to be used for research and analysis with aggregate data, such as summarizing characteristics of the payee population, investigating factors associated with misuse, and drawing samples for monitoring payees.

RECOMMENDATION 6.14 The Social Security Administration should establish mandatory protocols for payee replacement when misuse is suspected. When misuse or suspected misuse is the reason for a change of payee, staff should provide full documentation.

RECOMMENDATION 6.15 The Social Security Administration should redesign the Representative Payee System.

The committee suggests that SSA consider the following changes to the RPS:

(1) inclusion of all payees into the system;
(2) creation of data elements in the system with respect to a payee who is identified as a potential or suspected misuser;
(3) addition of data elements in the system for various types of violations by payees;
(4) addition of data elements in the system for relevant results of investigations by the Office of the Inspector General;
(5) inclusion of the Employer Identification Numbers of all organizational payees in the system;

(6) addition of a lump-sum indicator and amount to enable the local field office to better monitor how such money is spent for a specific beneficiary;

(7) easy access and use by all field office staff;

(8) streamlined linkage to annual accounting form data; and

(9) an easy-to-use interface that has undergone usability testing.

The committee suggests that SSA require entry of important data elements—standardized values that ensure consistency of responses across offices and support institutional analysis of the full population or special populations of payees. A new RPS should undergo usability testing to ensure that it effectively supports office staff in entering and updating the system. These improvements could be logical considerations under SSA's currently planned revision of the RPS.

RECOMMENDATION 6.16 The Social Security Administration should implement a process that regularly updates information in the Representative Payee System, both by field office staff and through the annual accounting form. The Social Security Administration should also implement a quality control program that periodically checks the integrity of the information in the Representative Payee System.

Finally, when there is change of payee the accounting process often falls by the wayside, particularly when the termination is abrupt and so there is a potential for the unaccounted use of funds.

CONCLUSION Whenever suspected cases of misuse are not subjected to a formal investigation but handled by use of the phrase "more suitable payee found," potential misusers are not held accountable for their actions; this approach may allow inappropriate reentry to payee status, and consequently future misuse.

CONCLUSION Lack of the required final accounting for terminations may cover up misuse, especially in cases in which a "more suitable payee" was found.

RECOMMENDATION 6.17 The Social Security Administration should revise the current regulations that require a final accounting whenever a payee is terminated to ensure, so far as practicable, that all funds are accounted for.

Effects of and Coordination with State Policies

State policies influence the administration of the Representative Payee Program in such areas as fees for service, training, accountability, monitoring, and oversight. There are significant state-to-state variations. For example, benefit funds are used to pay fees for representative payee services without regard for federal legislative limitations because of the way in which the SSA defers to state court oversight of guardianship and conservatorship financial reporting. State court guardianship and conservatorship programs operate totally independently from the SSA Representative Payee Program even though the program requires any beneficiary who has a guardian or conservator to also have a payee appointed by SSA. In addition, there is no coordination between SSA and state courts for the training of guardians, conservators, and payees regarding the filing of annual reports.

CONCLUSION Funds from both the Old Age, Survivors, and Disability Insurance Program and from the Supplemental Security Income Program are used to pay fees for representative payee services without regard for legislative limitations because of the way in which the Social Security Administration defers to state court oversight of guardianship and conservatorship financial reporting.

CONCLUSION State court guardianship and conservatorship programs operate totally independently from the Social Security Administration Representative Payee Program even though the program requires any beneficiary who has a guardian or conservator to also have a payee appointed by SSA. There is no coordination between SSA and state courts for the training of guardians, conservators, and payees or regarding filing annual reports.

RECOMMENDATION 6.18 The Social Security Administration should track state laws that require conservators or legal guardians of beneficiaries who need representative payees to undergo court monitoring and mandated training. In such states, the Social Security Administration should give preference to designating the guardians or conservators as the payees and seek to integrate or coordinate its payee training materials with the state-mandated training.

RECOMMENDATION 6.19 The Social Security Administration should begin an outreach program with state agencies to compile the laws and practices and study the differences in various states' regulation of assisted living, foster care, and other group homes.

The committee was impressed by the dedication of the SSA headquarters staff who administer the program, the claims representatives in the field offices, and the great majority of representative payees. The committee is confident that adoption of the 28 recommendations in this report will correct the deficiencies found in our review. Although each recommendation can be implemented or rejected on its own merits, the committee urges a comprehensive approach to their evaluation so that the benefits of each one is leveraged to enhance others.

1

Introduction:
The Representative Payee Program

he Social Security Administration (SSA) provides Old Age, Survivors, and Disability Insurance benefits (known as Social Security, OASDI, or Title II) and Supplemental Security Income benefits (known as SSI or Title XVI) to many of the most vulnerable members of society: the young, the elderly, and people with disabilities. In December 2006 almost 54 million people received almost $50.5 billion in benefits: under OASDI, to retired or disabled workers and spouses and children of retired, disabled, or deceased workers; under SSI, to disabled adults and children with limited income and resources and to people aged 65 and older with limited income and resources.

Congress recognized from the inception of these programs that some beneficiaries would need assistance in managing their benefits, and SSA's Representative Payee Program is designed to provide this assistance. A representative payee is an individual or organization that receives OASDI or SSI benefits for someone who cannot manage, or direct someone else to manage, his or her money. Representative payees are required to use the funds in the best interest of the beneficiaries. Currently, there are about 5.3 million representative payees serving more than 7 million beneficiaries.

SSA tries to select as a representative payee someone who wishes to help the beneficiary, and someone who can see the beneficiary often and who knows his or her needs. For that reason, if a beneficiary is living with someone who helps him or her, SSA usually selects that person to be the payee. In most cases, someone who knows the beneficiary asks SSA if she or he can be the payee. It may be a family member, a friend, a legal guardian, or a lawyer. Sometimes, however, social service agencies, nursing homes, medical, religious, or custodial institutions are selected as payees. SSA requires beneficiaries who have a state-court-appointed guardian to have a representative payee. Guardians have typically already undertaken similar fiduciary responsibilities and reporting obligations and are subject to liabilities for mismanagement of funds. Therefore, in such cases, they are usually selected as the representative payee. In some cases in which a payee cannot be found, the beneficiary becomes of age, or the beneficiary recovers from a temporary incapacity, the beneficiary is placed in direct pay status and in effect, becomes his or her own payee in the SSA administrative system.

BACKGROUND

On March 2, 2004, President Bush signed the Social Security Protection Act of 2004 (Public Law 108-203). Section 107 of the law requires the commissioner of Social Security to conduct a one-time survey to determine how payments to individual and organizational representative payees are being managed and used on behalf of the beneficiaries. The commissioner was directed to (1) assess the extent to which representative payees are not performing their duties in accordance with SSA standards for representative payee conduct, (2) learn whether the representative payment policies are practical and appropriate, (3) identify the types of representative payees that have the highest risk of misuse of benefits, and (4) find ways to reduce the risk of misuse of benefits and ways to better protect beneficiaries (see Boxes 1-1 and 1-2). The committee structured its tasks and this report to meet that charge. Although the committee's charge is nested within the larger realm of an ideal payee and what beneficiaries need and thus human and behavioral issues, the committee was not constituted and did not attempt to explore issues beyond the specific charge from SSA. Explicitly, the committee's charge did not include a broad assessment that might lead to recommendations that would require substantially new resources or fundamentally change the program. Rather, the committee's charge was directed to how the program is managed and for consideration of improved systems.

The legislative history of Section 107 ("Manager's Amendment," Congressional Record, December 9, 2003) indicates that the required survey shall assess the extent to which representative payees are failing to perform their duties as payees in accordance with SSA standards of payee conduct, including whether the funds are being used for the benefit of the beneficiary. It also indicates that, to the extent possible, the types of payees who have the highest risk of misuse of benefits should be identified, along with

---

BOX 1-1
Authorization for Study

SEC. 107. SURVEY OF USE OF PAYMENTS BY REPRESENTATIVE PAYEES. (a) IN GENERAL.—Section 1110 of the Social Security Act (42 U.S.C. 1310) is amended by adding at the end the following: "(c)(1) In addition to the amount otherwise appropriated in any other law to carry out subsection (a) for fiscal year 2004, up to $8,500,000 is authorized and appropriated and shall be used by the Commissioner of Social Security under this subsection for purposes of conducting a statistically valid survey to determine how payments made to individuals, organizations, and State or local government agencies that are representative payees for benefits paid under title II or XVI are being managed and used on behalf of the beneficiaries for whom such benefits are paid. "(2) Not later than 18 months after the date of enactment of this subsection, the Commissioner of Social Security shall submit a report on the survey conducted in accordance with paragraph (1) to the Committee on Ways and Means of the House of Representatives and the Committee on Finance of the Senate."

BOX 1-2
Definition of Misuse

"Misuse occurs in any case in which the representative payee receives payment under this title for the use and benefit of another person and converts such payment, or any part thereof, to a use other than for the use and benefit of such other person." Social Security Protection Act, §1631(a)(2)(A)(iv).

---

suggestions of ways to reduce those risks and better protect the beneficiaries. According to the Manager's Amendment, the survey should focus on representative payees who are not subject to triennial onsite review or other random review under SSA policy or law. Therefore, the groups to be included in the mandated study are individual representative payees who serve one or more but fewer than 15 beneficiaries and non-fee-for-service organizational payees who serve fewer than 50 beneficiaries.

THE COMMITTEE'S WORK

In September 2004 the National Academies accepted the charge from SSA to undertake the mandated study. The National Academies formed a volunteer, interdisciplinary committee of 11 members with expertise in survey methodology, program and process evaluation, SSA policies, and the experiences of disadvantaged people (see Appendix G for biographical sketches of committee members and staff). The committee held 20 meetings and phone conferences between April 2005 and April 2007 to examine existing reports and legislation relevant to the Representative Payee Program and to analyze SSA administrative records, deliberate, and write this report. As part of its work, the committee developed and, after a competitive process, awarded a contract to Westat to conduct a survey of representative payees and beneficiaries. The sample design selected 5,098 representative payees and paired 2,543 beneficiaries with their representative payees using a complex, multistage sampling plan. Readers of this report can see the extensive methodology in Appendix A, available online, including the questionnaires, in order to understand how the survey was approached and implemented.

Committee members visited 13 field offices, a regional office, a data operation center, and a payment center. Seven of the field offices were located in central cities of large metropolitan areas, one office was in a suburban area, and five were urban offices serving primarily rural areas. The offices were in eight different states (Arizona, California, Illinois, Iowa, Kentucky, Maryland, New York, and Virginia) and the District of Columbia. Numerous hours were spent interviewing SSA staff during these visits to obtain information on their experiences and perspectives on the Representative Payee Program. As a result of the committee's visits and analyses, the committee submitted two lists of questions to SSA: the questions and SSA's responses are in Appendixes B and C.

Committee members also looked in depth at the administrative record system of the SSA, including a review of the folders of known misuser cases (see Appendix D), an analysis of lump-sum payments[1] to beneficiaries and a review of the annual accounting form and how it is processed (see Chapter 4), and a review of the interactive Representative Payee System (RPS) (see Chapter 6).

To specifically study the issue of misuse, the committee worked with a forensic auditor and Westat to develop a reinterview process of a small number of select respondents from the original survey of representative payees. The committee hypothesized potential characteristics of misusers in selecting the participants for reinterview. This follow-on activity allowed the interviewers to probe specific issues about how representative payees managed the funds of beneficiaries (see also Chapter 5).

---

[1] A lump-sum payment to a beneficiary can be several thousands of dollars. Such a payment may arise when benefits have been withheld pending review of a disability claim or when benefits have been withheld pending the appointment of a representative payee.

The committee used all the available sources of information (site visits, administrative data, and surveys) to develop, understand, and assess information. The committee did not give undue emphasis (recognizing the limitations and shortcomings of each source) to any one source of information. The committee found that all the sources of information added value in answering the charge from SSA.

The rest of this chapter describes the Representative Payee Program, and Chapter 2 describes the target universe of representative payees and beneficiaries. The subsequent chapters then address the four key areas specified by the commissioner of SSA: Chapter 3 covers representative payee performance; Chapter 4 looks at the characteristics of misuse and misusers; Chapter 5 covers ways to reduce the risk of misuse of benefits and to better protect beneficiaries; and Chapter 6 evaluates SSA policies and procedures for the Representative Payee Program. The committee's conclusions and recommendations are included in Chapters 3 through 6. As with any study of a program of this magnitude, the committee's conclusions and recommendations should be understood in the context of some deficits in practical information, such as accurate and timely names and addresses of payees, the inherent missing or errant data in huge administrative files, and the qualitative nature of interviewing SSA staff.

PROGRAM OVERVIEW

The Social Security Act Amendments of 1939 (ch. 666, § 205(j), 53 Stat. 1360, 1371) gave broad authority to the SSA to appoint representative payees for those beneficiaries who were deemed incapable of managing or directing the management of their benefits. SSA's current statutory authority for the Representative Payee Program is found at 42 U.S.C. §§ 405(j), 1007, & 1383(a)(2).

Children under age 18 are assumed to be unable to manage their funds. Beneficiaries aged 18 or older are generally assumed to be able to manage their funds unless convincing evidence is provided to rebut this assumption. In determining whether an individual is unable to manage or direct the management of benefit payments, SSA considers evidence from medical, legal, and lay sources. Other than children, most beneficiaries with payees are those whose mental or physical conditions prevent them from acting on their own behalf and those persons adjudged incapable under state guardianship laws.

The size of the Representative Payee Program has grown along with the expansion of SSA's responsibilities. Originally, SSA benefits were primarily provided to eligible retirees and their families. When SSA began issuing disability benefits in 1956 and SSI payments in 1974, the demographics of its beneficiary population changed from retirees and their families to include

TABLE 1-1 Number of People Receiving OASDI, SSI, or Both, March 2007 (in thousands)

Type of Beneficiary	Total	OASDI Only	SSI Only	Both OASDI/SSI
All beneficiaries	54,167	46,880	4,727	2,529
Aged 65 or older	35,541	33,531	864	1,147
Disabled, under age 65[a]	11,446	6,170	3,864	1,412
Other[b]	7,180	7,180	N/A	N/A

NOTES: Data are for the end of the specified month. Only beneficiaries in current-payment status are included. Due to rounding, numbers may not add to the total.

[a]Includes children receiving SSI on the basis of their own disability.

[b]Beneficiaries who are neither aged nor disabled (for example, early retirees, young survivors).

SOURCE: Data from Social Security Administration, Master Beneficiary Record, 100 percent data and Social Security Administration, Supplemental Security Record, 100 percent data.

persons with disabilities, blind individuals, and the elderly with limited income and resources. In March 2007, SSA served over 54 million beneficiaries (Table 1-1) with over 7 million (Table 1-2) having representative payees. Some payees serve more than one beneficiary, some payees are not entered into the administrative database, and beneficiaries are constantly changing payees, so SSA does not have a current, accurate count of payees. In its December 2005 statistical report SSA's Office of Policy estimated there are 5.3 million representative payees. For almost 64 percent of beneficiaries with a representative payee, their payee is a natural, adoptive, or stepparent (Table 1-2).

TABLE 1-2 Number and Percent of Beneficiaries by Representative Payee Type (December 2005)

Type of Representative Payee	Number	Percent
Spouse	256,902	3.6
Parent	4,528,668	63.7
Adult child	300,737	4.2
Other relative	944,100	13.3
Institution	562,210	7.9
Agency/financial organization	250,694	3.5
Other	264,876	3.7
Total	7,108,187	100.0

SOURCE: Data from Social Security Administration Representative Payee Program staff.

Of those beneficiaries with representative payees, approximately 64 percent receive OASDI benefits only (payments for retired and disabled workers, their survivors and dependents), approximately 28 percent receive SSI only (persons with limited income who are disabled, blind, or elderly) and approximately 8 percent receive benefits from both programs (Table 1-3).

Approximately 4.1 million beneficiaries who have a representative payee are minor children. Of those, about 75 percent receive OASDI ben-

TABLE 1-3 Number and Percent of Beneficiaries by Representative Payee Type Receiving OASDI, SSI, or Both (December 2005)

Payee Type	Type of Benefit	Number	Percent
Spouse			
	OASDI only	197,500	76.9
	SSI only	44,212	17.2
	Both OASDI/SSI	15,190	5.9
Parent			
	OASDI only	3,063,310	67.7
	SSI only	1,228,745	27.1
	Both OASDI/SSI	236,613	5.2
Adult child			
	OASDI only	204,390	68.0
	SSI only	63,451	21.1
	Both OASDI/SSI	32,896	10.9
Other relative			
	OASDI only	492,940	52.2
	SSI only	335,780	35.6
	Both OASDI/SSI	115,380	12.2
Institution			
	OASDI only	358,620	63.8
	SSI only	135,305	24.1
	Both OASDI/SSI	68,285	12.1
Agency/financial organization			
	OASDI only	105,680	42.2
	SSI only	91,973	36.7
	Both OASDI/SSI	53,041	21.2
Other			
	OASDI only	114,410	43.2
	SSI only	107,532	40.6
	Both OASDI/SSI	42,934	16.2
Total			
	OASDI only	4,536,850	63.8
	SSI only	2,006,998	28.2
	Both OASDI/SSI	564,339	7.9

SOURCE: Data from Social Security Administration Representative Payee Program staff.

TABLE 1-4 Number and Percent of People Under 18 with a Representative Payee Receiving OASDI, SSI, or Both (December 2005)

Type of Benefit	Number	Percent
OASDI only	3,050,570	74.7
SSI only	961,556	23.5
Both OASDI/SSI	74,095	1.8
Total	4,086,221	100.0

SOURCE: Data from Social Security Administration Representative Payee Program staff.

efits only, about 24 percent receive SSI only, and about 2 percent receive concurrent benefits (Table 1-4).

Selection of Representative Payees

SSA tries to select as a representative payee someone who knows and wishes to help the beneficiary. Some of the factors SSA considers in representative payee selection are the following: familial or custodial relationship with the beneficiary; demonstration of concern for the beneficiary; legal authority to act on behalf of the beneficiary; and a favorable record of prior service as a representative payee (or lack of any unfavorable record such as suspicion of mismanagement of beneficiary funds). SSA also reviews whether the potential payee is a creditor of the beneficiary and whether the payee has a criminal history. If possible, SSA seeks the beneficiary's approval of the selected payee. Evaluating the criteria for selecting a payee was not part of the charge to the committee. The committee observed in site visits that claims representatives have a very difficult task to choose a representative payee in applying criteria among blood relationships (especially in the case of minor children), the competency of a representative payee, the living arrangement of the beneficiary, or simply a very caring person. There is no one perfect criterion. In many cases, SSA does not have the luxury of choosing among potential representative payees and is very satisfied to find one, even a less than optimal, willing payee.

The payee selection process is initiated by written application generated by family members, Disability Determination Services,[2] the courts, or other concerned individuals who perceive the need for a payee or directly by SSA field staff (typically claims representatives) when the agency has

---

[2]Most disability claims are initially processed through a network of local SSA field offices and state agencies (usually called Disability Determination Services or DDSs).

received information indicating such a need. The SSA investigates and decides whether to appoint a payee based on its regulations, 20 C.F.R. §§ 404.2001-.2065, 416.601-.732 (2006), and its Program Operations Manual System,[3] (GN 00502.105, Citations: 20 C.F.R., Sections 404.2021, 416.621 and GN 00502.130).

Individuals who serve as representative payees include the spouse, parent, stepparent, grandparent, adult child, and other relatives of the payee, friends, legal guardians, attorneys, and nonrelatives. Organizational payees may be various types of community-based organizations, institutions, government agencies, and financial organizations. Examples of organizational payees include Department of Veterans Affairs hospitals, state psychiatric institutions, foster homes, and community social service groups. Some organizations are set up to handle specific types of beneficiaries such as those beneficiaries with severe head injuries.

Individual payees cannot collect fees, but an approved fee-for-service organization may collect a fee from the monthly SSA benefit payments if the organization meets certain conditions. To qualify as a fee-for-service payee, the organization must:

- be a state or local government agency, or
- be a community-based, nonprofit social service agency, bonded and licensed in the state in which it serves as a representative payee, and
- regularly provide representative payee services to at least five beneficiaries, and
- not be a creditor of the beneficiary.

Fee-for-service organizations are subject to regular reviews by SSA. The fee is limited to not more than 10 percent of the monthly benefit or $33 whichever is lower. An organization may not collect a fee if the organization is receiving compensation from another source, such as court or guardianship fees.

SSA requires beneficiaries who have a state-court-appointed lawyer, conservator, or guardian to also have a representative payee. In such cases, these officials, who may have already undertaken similar fiduciary responsibilities and reporting obligations and who are subject to liabilities for mismanagement of funds, are typically asked to serve as the payee.

---

[3]SSA employees refer to the operations manual to administer the OASDI and SSI programs. (See https://s044a90.ssa.gov/apps10/poms.nsf/aboutpoms [June 2007].)

Duties of Representative Payees

According to SSA regulations (20 C.F.R. §§ 404.2035 and 416.635), the charge to representative payees is to use the funds in the best interest of the beneficiaries; that is, to pay for the current and foreseeable needs of beneficiaries. The required duties of representative payees are to

- determine the beneficiary's needs and use his or her payments to meet those needs;
- save any money left after meeting the beneficiary's current needs in an interest bearing account or savings bonds for the beneficiary's future needs;
- report to SSA any changes or events which could affect the beneficiary's eligibility for benefits or payment amount (such as time in prison, earning of significant wages, etc.);
- keep records of all payments received and how they are spent and/or saved;
- provide benefit information to social service agencies or medical facilities that serve the beneficiary;
- help the beneficiary obtain medical treatment when necessary;
- notify SSA of any changes in the payee's circumstances that would affect his or her performance or continuing as payee;
- complete and submit required written reports accounting for the use of funds; and
- return to SSA any payments to which the beneficiary is not entitled (such as payments made while beneficiary was in prison, death of the beneficiary, etc.).

Among the duties of a payee is the proper handling and accounting for lump-sum payments. Most lump sums are retroactive benefits, i.e., benefits accrued prior to an award. In the Representative Payee Program, lump sums might also accrue while payment is suspended for selection of a new payee. Lump sums are usually paid as a one-time payment if the amount combined with the monthly benefit is less than $4,000. If the combined sum is more than $4,000, the accumulated or conserved funds must be paid in installments.

Lump sums can be large, especially if there is a delay in determining disability status. During the time period 2000 to 2004, the largest lump sum paid to any payee was $125,000. This amount was received by an organization with 29 beneficiaries. The largest payment to an individual payee was in the amount of $67,130.

Since 1974, the SSI children's program has provided monthly benefits for children with disabilities and blind under age 18. Children who receive SSI benefits because of a disability may have lump-sum retroactive awards that go into the thousands of dollars. SSA has clear instructions on the proper way to deal with such amounts (20 C.F.R. § 416.625). The payments must be paid directly into a dedicated account in a financial institution (20 C.F.R. § 416.546). The account must be maintained separately from any other savings or checking account set up for the child. The payee may use the funds only for certain expenses, primarily those related to the child's disability (such as medical treatment, education, job training, rehabilitation, special equipment, or house modifications). The payee must be able to explain how the item or the service benefits the child and must keep records and receipts of all deposits to and expenditures from dedicated accounts.

Monitoring and Support of Representative Payees

Given that it is not feasible to determine directly from all beneficiaries whether their representative payees are acting in their best interests, SSA has developed policies that focus on selecting individuals who seem likely to be able to meet the program goals and policies designed to monitor the ongoing performance of the representative payees.

All payee applicants are supposed to be interviewed face to face by SSA field office staff, although this does not always occur, and some organizations are exempt from the face-to-face requirement. Whether the interview is in person or by telephone, the applicant's identity is checked and other information gathered. Only during the payee application process is there any attempt to establish an understanding of the responsibilities of the payeeship and to evaluate if the payee is capable of properly executing them. SSA really does not know, for example, if a prospective payee is even capable of managing his or her own finances.

Research has shown there are potential problems in the selection process (Elbogen et al., 2005a, 2005b). For example, research on beneficiaries with psychiatric disabilities has shown that payeeship, if misunderstood or misapplied, can be used coercively, foster dependency, and lead to conflict and even violence.

According to the survey and site visits, after a payee starts managing the Social Security funds of a beneficiary, there is little in the way of continuing, proactive support by SSA for the payee. Most payees have very little, if any, subsequent contact with SSA about their duties and responsibilities. There is no refresher training or follow-up meeting to assess payee behavior, quality of performance, or answer questions. A brochure that reminds the payee of their duties is included with the annual accounting form. SSA does maintain a website with information about payee duties. But in general, once a payee starts his/her service, there is no support or interaction between SSA and the payee unless the payee initiates the contact.

The current ongoing monitoring program requires all payees to file a two-page annual accounting report that is mailed to them. SSA tries to use the report to monitor how the payee spent or saved the benefits and to identify situations where the beneficiary may no longer need a payee or where the payee is no longer suitable. Approximately one-twelfth of the representative payees are mailed an accounting form each month from SSA's Wilkes-Barre, PA, Data Operations Center. There are three main questions regarding how funds were saved or spent on food, housing, education, clothing, etc. during the accounting period. On the form SSA tells the payee the total funds received. Clerks manually review each form and decide if follow-up is necessary. If the amounts given for the three questions come close to adding up to the stated total, the report is deemed acceptable. In some cases, after a phone call from SSA staff in the processing center to the payee has not resolved apparent issues, the form is sent to a local office for follow-up. If the payee does not respond in a timely manner to the local office, beneficiary checks can be held until the payee appears at the local office to clear up any deficiencies.

In addition to the annual accounting form, certain payees are also subject to triennial site reviews which are conducted through a face-to-face meeting. These reviews are conducted on "volume payees" (organizational payees serving 50 or more beneficiaries or individuals who are serving 15 or more beneficiaries), all fee-for-service payees, and state mental institution payees (none of these types of payees are part of this current study). SSA also conducts site reviews in response to events that raise concerns about a particular payee's suitability or performance of duties.

Representative Payee System

A critical tool in the support and monitoring of representative payees is the RPS. The RPS is a database system used to enter and maintain information about representative payees and the beneficiaries they serve. The RPS is mandated by statute (42 U.S.C. § 405(j)(2)(B)(ii)) that requires SSA to establish and maintain a centralized file, readily retrievable by SSA offices. The RPS contains the names of almost all payees, their beneficiaries, type of benefit, dates of service, contact information, and information such as whether their status has been revoked by reason of misuse of funds or whether they did not serve their beneficiary according to SSA rules. This is a critical system in administering the Representative Payee Program, and is heavily used by SSA offices to document payee relationships and evaluate potential payees for service. However, the RPS is not easy or efficient to use for statistical analysis purposes.

CONCLUSION

The Representative Payee Program is an important program that by its sheer size and shifting needs is difficult to administer, support, and monitor. As with many programs of this magnitude, SSA is limited by resources in its efforts to select, educate, support, and monitor representative payees. Representative payee policies and practices are oriented towards careful selection and monitoring of representative payees as effective ways to assure that the best interests of the beneficiaries are being served. In a program that involves over 5.3 million representative payees, more than 7 million beneficiaries, and almost $4 billion in monthly benefit payments, the selection and monitoring of the representative payees is a daunting task. According to SSA, for some beneficiaries and in some geographic areas, it is difficult to even find individuals or organizations willing to serve.

2

Representative Payees and Their Beneficiaries

The committee's survey of representative payees and beneficiaries provided information on the populations that are the focus of this report. The first section of this chapter describes the survey selection criteria. The second section reports on the characteristics associated with payeeship, including time as a payee and beneficiary volume, demographic characteristics, educational attainment, measures of financial situation (income and source of income), and other characteristics that might be associated with payee performance, such as stability in community (frequency of move) and criminal background. The third section looks at some of the characteristics of beneficiaries. (For details of the survey of payees and beneficiaries, see Appendix A.)

SURVEY SELECTION CRITERIA

As described in Chapter 1, the population for the committee's survey was individual representative payees serving fewer than 15 beneficiaries and non-fee-for-service organizational payees serving fewer than 50 beneficiaries. Our survey covered all such payees serving on January 1, 2006. To make the survey manageable and meaningful, we restricted our population of payees to those who satisfied all of the following conditions: (1) resided in the 48 contiguous states, (2) had one or more current beneficiaries, (3) had an address with a valid state and county code on the administrative records, and (4) managed funds of more than $50 each month for one or more beneficiaries.

The population of beneficiaries was restricted to all persons who on January 1, 2006, were (1) managed by a survey eligible payee, (2) age 14 and older, and (3) received more than $50 in benefits each month. If a payee served more than one beneficiary, a beneficiary was randomly selected to form a dyad with the payee. This strategy allowed questions to be asked of a payee in relation to a specific beneficiary.

Because over 4 million of the 7 million beneficiaries who need payees are less than 18 years of age, a strictly random sample would contain a very large number of payees whose beneficiaries are children. Thus, the committee identified specific domains of study to ensure that the survey sample would include a wide variety of other types of payees beyond parents of beneficiary children.

With these specifications, the committee commissioned Westat to select a multistage sample of 5,098 payees and attach a beneficiary to 2,543 of those payees. Table 2-1 lists the domains we selected and the counts of payees and beneficiaries included in the study. The domains represent about 3.5 million payees serving 4.6 million beneficiaries of which 2.8 million are age 14 or older.

If a beneficiary who was selected to be interviewed was deemed to be incapable of participating in the survey, a proxy was used.[1] The responses to survey questions collected from proxies were compared to the responses obtained from beneficiaries. The review of the data revealed that differences between the two were either not statistically significant or differences were no greater than three percentage points. Thus, a breakdown of these small differences in the survey results is not shown.

CHARACTERISTICS OF REPRESENTATIVE PAYEES

Over time, an individual payee rarely served for more than a few beneficiaries. An estimated 65.8 percent (1.2)[2] of representative payees had served only one beneficiary in their lifetime, 30.3 percent (1.1) had served two or three beneficiaries, and 3.9 percent (0.5) had served four or more

---

[1] A proxy is a capable adult who is close enough to the beneficiary's daily life that he or she can complete an interview on the beneficiary's behalf. In this survey, neither the representative payee nor the spouse of the payee could serve as a beneficiary's proxy. If the beneficiary lived in group quarters, a caregiver, caseworker, or nurse was preferred to a family member.

[2] For all estimates in this chapter, the number in parentheses following the estimate is the standard error of the estimate. As a general rule, users can approximate a 95-percent confidence interval for the estimate by adding and subtracting two standard errors to the estimate. When two estimates have confidence intervals that overlap, the two estimates are not statistically different at the .05 level of significance.

TABLE 2-1 Domains and Number of Representative Payees and Beneficiary Cases

Repesentative Payees by Type of Beneficiaries	Number			
	Unweighted		Weighted	
	Payees	Beneficiaries[a]	Payees	Beneficiaries
With child beneficiaries (14-17)	884	438	134,583	127,749
With beneficiaries aged 18-64	2,601	1,446	1,407,417	1,637,553
With beneficiaries aged 65+	1,032	578	332,900	403,593
With unrelated beneficiaries	1,465	519	2,502,119	1,381,854
With only one unrelated beneficiary	858	451	28,959	63,575
Who are either an adult child or another relative of the beneficiary	1,390	716	1,083,520	1,068,158
Who are a parent of the beneficiary	1,501	701	2,969,738	2,047,260
Who live with the beneficiary or unknown living arrangement	3,622	1,775	4,000,197	3,086,899
Who do not live with beneficiary	1,011	531	216,603	219,843
With a beneficiary who receives SSI	2,485	786	1,622,642	1,330,480
With a beneficiary who receives OASDI	3,133	1,618	2,283,017	1,963,247
Who are individuals	4,633	2,306	4,216,800	3,306,742
Who are organizations	465	237	25,636	116,258

NOTE: The domains include 707,573 representative payees and 584,386 beneficiaries whom the committee excluded from the survey because (1) their monthly benefit was less than $50, (2) benefit issuance date was missing, (3) there was a disagreement in files about beneficiary name, (4) beneficiary had two payees, or (5) field-determined ineligibility; see Appendix A for details.

[a]Aged 14 or older.

SOURCE: Data from the national survey of representative payees and beneficiaries conducted for the National Academies Committee on Social Security Representative Payees (2006).

beneficiaries. From the other perspective, most beneficiaries, 81.1 percent (1.2) had had only one payee in their lifetimes, 13.4 percent (1.3) had two, and 4.7 percent (0.7) had had three or more.

As expected, organizational payees had a higher volume. An estimated 11.1 percent (4.0) of the organizations in our sample had served only one beneficiary, 5.7 percent (1.8) had served two or three, and 83.2 percent (4.4) had served more than four. It should be kept in mind that an organization may also have had other clients in their care for whom they did not serve as a representative payee.

The average length of service to all current and prior beneficiaries for the representative payees was between 4 and 5 years, but a few payees had been involved with the program for close to 40 years. Table 2-2 shows the length of time as payee by type of payee. In the sampled universe, organizations had served longer than individuals. An estimated 31.2 percent (1.6) of individual payees had served for more than 10 years compared with 53.2 percent (4.2) of organizations.

In the survey, close to 50 percent of the payees managed only Old Age, Survivors, and Disability Insurance (OASDI) benefits, 45 percent (1.4) managed only Supplemental Security Income (SSI) benefits, and the rest managed both OASDI and SSI monthly benefits.

The committee was interested in other payee characteristics, which were included in the survey. One such characteristic was the fraction of representative payees who had access to the Internet. It is important to determine whether the Social Security Administration (SSA) might be able to rely on the Internet for monitoring payee performance or to provide additional tools to assist representative payees. An estimated 61.3 percent (1.4) of representative payees had access to the Internet; of that group, an estimated 84 percent (1.1) had access to the Internet at home, 35.9 percent (2.1) at work, 50.2 percent (2.6) at the library, and 19.4 percent (1.5) somewhere else.

TABLE 2-2 Length of Time as a Representative Payee by Type of Payee

Time	Individual Percent	Organization Percent
Less than 2 years	9.2 (1.0)	6.8 (2.2)
2 to 5 years	29.2 (1.8)	20.5 (3.4)
5 to 10 years	30.4 (1.4)	19.6 (3.7)
10 or more years	31.2 (1.6)	53.2 (4.2)
Total	100.0	100.0

NOTE: Numbers in parentheses are the standard errors of the estimates.

SOURCE: Data from the national survey of representative payees and beneficiaries conducted for the National Academies Committee on Social Security Representative Payees (2006).

TABLE 2-3 Representative Payee and Beneficiary Race and Ethnicity

Race and Ethnicity[a]	Payees	Beneficiaries
Hispanic	13.4 (2.5)	16.2 (2.8)
White	67.7 (3.6)	66.7 (4.2)
Black or African American	25.3 (3.7)	27.0 (4.3)
American Indian or Alaska Native	8.1 (1.1)	9.5 (1.0)
Asian	2.7 (0.5)	2.5 (0.5)
Native Hawaiian or Pacific Islander	0.5 (0.2)	1.2 (0.4)

NOTE: Numbers in parentheses are the standard errors of the estimates.

[a]A respondent can be in more than one race category.

SOURCE: Data from the national survey of representative payees and beneficiaries conducted for the National Academies Committee on Social Security Representative Payees (2006).

The payees in the survey were most likely to be under age 50 and female. About 13 percent (2.5) were of Hispanic or Latino origin, which is similar to the national population. About 68 percent (3.6) identified themselves as white, 25 percent (3.7) as black or African American, 8 percent (1.1) as American Indian or Alaska Native, 2.7 percent (0.5) as Asian, and 0.5 percent (0.2) as Native Hawaiian or other Pacific Islander (see Table 2-3). For comparison, nationally, in 2003, for people reporting one race alone, 78 percent identified themselves as white, 12 percent as black or African American, 1 percent as American Indian and Alaska Native; 4 percent as Asian; less than 0.5 percent as Native Hawaiian or other Pacific Islander, and 5 percent as "some other race" (U.S. Census Bureau, 2003).

The educational level of representative payees is somewhat lower than the national average with 76.6 percent having graduated from high school and only 13.3 percent having a college degree. For comparison, in 2003, nationally, 84 percent of people 25 years and older had at least graduated from high school and 27 percent had a bachelor's degree or higher (U.S. Census Bureau, 2003).

The survey asked several questions about characteristics that are important in selecting a representative payee or that may be related to the risk that a representative payee will misuse beneficiary funds. These questions included the representative payee's income, sources of income, frequency of residential changes, and criminal record.

The representative payees reported personal incomes significantly lower than the national average. More than 25 percent (1.8) of the payees reported annual 2005 incomes of less than $5,000, and close to 33 percent reported between $5,000 and $15,000. Nationwide in 2003, only 8 percent of the population reported individual incomes below $5,000, and only 17 percent reported between $5,000 and $15,000 (U.S. Census Bureau, 2004). Similarly, of representative payees, 33 percent had incomes between $15,000 and $50,000, compared with 46 percent in the nation. In the survey, very few payees reported incomes of more than $50,000, compared with 27 percent in the nation. (It should be noted that the reporting of individual incomes may not reflect potentially much larger household incomes.)

For income, 9 percent (0.7) of representatives were self-employed, and almost 50 percent (2.2) said they received income from employers. Interestingly, close to 50 percent received their own OASDI benefits, SSI benefits, benefits from the U.S. Department of Veterans Affairs, or a pension. And more than 17 percent received income from the Temporary Assistance to Needy Families (TANF) Program or some other assistance program.

An estimated 28.4 percent (1.5) of payees had moved in the previous 2 years, with 19.6 percent (1.1) moving once and 6.7 percent (0.6) moving twice. In comparison, nationally in 2003, about 15 percent of the population reported moving in the past year (U.S. Census Bureau, 2004). Since the survey asked about the payees' moving in the past 2 years and the Census Bureau data are for 1 year, it is difficult to assess whether the mobility of payees differs from that of the general population. An estimated 14.9 percent (2.3) of payees in the survey used a mailing address that was different from their home address, suggesting that they either picked up mail at a post office box or had a business address.

Some payees had criminal backgrounds, but in that regard they do not appear to be different from the national population. Nationally in 2001, an estimated 2.7 percent of adults in the United States had served time in prison (Bureau of Justice Statistics, 2004). In the survey, an estimated 3.3 percent (0.5) of the payees reported that they had been convicted of a felony in the past, and 2.4 percent (0.4) had served time in prison.

A small fraction of payees, 1.1 percent (0.3), had been treated in the past 5 years for alcohol or drug problems. A few, 0.3 percent (0.1), said they felt they needed treatment, and a small fraction, 0.2 percent (0.1), reported they were told they needed treatment.

CHARACTERISTICS OF BENEFICIARIES

Many representative payees serve because a beneficiary is a minor or incapacitated in some way. In the survey, about 35 percent of the beneficiaries were under the age of 18, about 53 percent were between 18 and 64, and about 12 percent were 65 years or older. In contrast with the payees, who were mostly female, the beneficiaries were more likely to be male (53.9 percent).

The beneficiaries were similar to the payees with regard to race and ethnicity, but different from the national distribution. The beneficiaries were predominantly white, 66.7 percent (4.2). About 27 percent (4.2) identified themselves as black or African American, 9.5 percent (1.0) as American Indian or Alaska Native, 2.5 percent (0.5) as Asian, and 1.2 percent (0.4) as Native Hawaiian or other Pacific Islander. When asked about Hispanic origin, 16.2 percent (2.8) identified themselves as being of Hispanic or Latino descent (Table 2-3).

The survey asked beneficiaries what language they used to talk with their payees. About 91.6 percent (1.2) said English, 5.9 percent (1.1) said Spanish, and 2.6 percent (0.6) said some other language. The beneficiaries reported that they communicated well in the chosen language.

Severe impairment prevented schooling for 3.4 percent (0.7) of the beneficiaries and, as expected, the overall educational attainment level of the beneficiaries was different from that of the payees. Most beneficiaries were still in school or had completed less than high school, 62.2 percent (2.2).[3] About 25 percent (2.0) had completed high school, and 2.3 percent (0.4) had a vocational or trade school education. Some beneficiaries had higher education: close to 5 percent (0.7) had some college or an associate degree, 1.8 percent (0.4) had a 4-year college degree, and a few, 0.6 percent (0.2), reported graduate school as the highest level of completed education.

Of beneficiaries who reported that they are still in school or participating in a training program, around 10 percent (2.0) said they were in a special program addressing activities of daily living. Of those for whom compensated work was possible, 14.5 percent (2.3) said they had worked full time (more than 35 hours per week) in the past 30 days.

During the past 5 years 4.6 percent (1.0) of beneficiaries had received treatment for alcohol or other drug problems. More than 2 percent (0.7) said that during the past 5 years they had been told that they needed treatment for alcohol or other drug problems, but none of the beneficiaries actually felt they needed treatment according to the survey.

SSA prefers a representative payee to be closely linked with the beneficiary. In the survey, almost all the beneficiaries knew their payees before they became the payees and more than 90 percent were related (see Table 2-4). Almost three-quarters of the beneficiaries lived with their representative payees, 72.2 percent (1.5). More than 50 percent lived in their own apartments or homes, and more than 25 percent lived in other people's apartments or homes. The rest lived in various kinds of group living situations (see Table 2-5). An estimated 83.1 percent (1.4) had been in their current living arrangement for more than 12 months, 7.4 percent (1.0) for 6-12 months, and 9.6 percent (1.0) for less than 6 months.

---

[3]The response category to the educational attainment question lumped "still in school" with completed schooling. A cross-tabulation of the educational attainment variable by beneficiary age shows that in our sample, 44.5 percent of the 62.2 percent are 18 or older.

TABLE 2-4 Beneficiary Relationship to the Payee

Representative Payee	Percent
Parent	61.2 (0.5)
Other relative	22.6 (0.7)
Adult child	8.3 (0.7)
Nonrelative	5.2 (0.1)
Organization	2.7 (0.1)
Total	100.0

NOTE: Numbers in parentheses are the standard errors of the estimates.

SOURCE: Data from the national survey of representative payees and beneficiaries conducted for the National Academies Committee on Social Security Representative Payees (2006).

TABLE 2-5 Beneficiary Residence

Usual Place of Residence[a]	Percent
Own apartment or home	61.9 (2.4)
Someone else's apartment or home	28.1 (2.1)
Group home	2.5 (0.3)
Residence for senior citizens	1.7 (0.3)
Nursing home	4.6 (0.4)
Long-term care hospital or related institution	0.5 (0.2)
Facility for persons with mental retardation or physical disability	0.4 (0.2)
Somewhere else	0.3 (0.1)
Total	100.0

NOTE: Numbers in parentheses are the standard errors of the estimates.

[a]Information on residence is from the beneficiaries.

SOURCE: Data from the national survey of representative payees and beneficiaries conducted for the National Academies Committee on Social Security Representative Payees (2006).

3

Performance of Representative Payees

This chapter addresses the first of the four specific items in the committee's charge: "assess the extent to which representative payees are not performing their duties in accordance with SSA standards for representative payee conduct." The findings in this chapter come from the committee's survey (see Appendix A) and 13 site visits to Social Security Administration (SSA) field and regional offices.

The findings in this chapter are presented in six substantive sections: (1) contact and communication between representative payees and beneficiaries, (2) the duties and responsibilities of payees, (3) communication with SSA, (4) handling beneficiaries' funds, (5) payee and beneficiary perceptions of how beneficiaries' needs are met, and (6) overall satisfaction and agreement between payees and beneficiaries. The final section presents the committee's conclusions and recommendations on representative payee performance.

CONTACT AND COMMUNICATION BETWEEN REPRESENTATIVE PAYEES AND BENEFICIARIES

The frequency and quality of communication between payees and beneficiaries are important for ensuring that beneficiaries' needs are being met: it is a key indicator of the degree to which payees are performing their duties.

Frequency of Communication

During the committee's site visits, SSA staff stressed that payees should be in frequent contact with their beneficiaries, and the survey results show such contact. The vast majority of payees, 86.1 percent (1.5),[1] reported being in touch with their beneficiaries once a week or more. About one in nine, 11.6 percent (1.3), reported monthly contact with their beneficiary; 2 percent (0.4) indicated contact once every few months; and a very small number, 0.4 percent (0.1), reported never having contact with their beneficiaries. When payees who were not living with their beneficiaries had contact, 78.6 percent (2.6) met in person and 19.7 percent (2.6) talked by telephone.

At the time of the interview, 70.2 percent (2.5) of payees reported their most recent contact with their beneficiary had been within the last week, and 16.8 percent (1.7) reported contact within the past month. Smaller proportions reported their most recent contact to have been 2-3 months ago, 5.1 percent (0.9), 4-6 months ago, 4.4 percent (0.7), 7-12 months ago, 2.5 percent (0.5), or more than 12 months ago, 1.0 percent (0.3).

Most beneficiaries who were not living with their payees also reported having weekly, 84.0 percent (2.0), or monthly, 10.5 percent (1.8), contact. A small number of beneficiaries indicated less frequent contact, 3.5 percent (0.8), or no contact, 2.0 percent (0.7). Consistent with payees' reports, beneficiaries also indicated their most common form of contact was in person, 77.8 percent (2.3), or by telephone, 21.6 percent (2.3). In addition, a majority reported their most recent contact with their payees had been within the past week, 79.6 percent (2.5), or the past month, 15.3 percent (2.5). For the rest, beneficiaries said that their most recent contact with their payees was 2-3 months ago, 2.9 percent (0.6), 4-6 months ago, 1.1 percent (0.3), 7-12 months ago, 0.8 percent (0.3), or more than 12 months ago, 0.4 percent (0.3). During the committee's site visits, SSA staff also cited a few instances in which beneficiaries had little or no communication with their payees, and said that a small number of beneficiaries had no idea where their payees were located.

Discussions of Beneficiary Needs

Table 3-1 presents the survey findings on payee and beneficiary reports of discussions during the past year on beneficiary needs. Most payees

---

[1]For all estimates in this chapter, the number in parentheses following the estimate is the standard error of the estimate. As a general rule, users can approximate a 95-percent confidence interval for the estimate by adding and subtracting two standard errors to the estimate. When two estimates have confidence intervals that overlap, the two estimates are not statistically different at the .05 level of significance.

TABLE 3-1 Representative Payee and Beneficiary Reports of Discussions of Beneficiary Needs During the Past Year (in percentage)

Beneficiary Needs	Payees Saying Yes	Beneficiaries Saying Yes
Housing, including rent, mortgage, repairs, utilities, furnishing, and needing to move to another place	44.0 (2.0)[a]	47.9 (2.1)[b]
Food and diet	68.0 (1.8)[c]	61.0 (2.1)[b]
Clothing, including buying and taking care of clothes	81.7 (1.1)[c]	69.5 (2.3)[b]
Medical care, including doctor, dentist, mental health, and vision services as well as medications	78.6 (1.3)[c]	73.4 (2.1)[d]

NOTE: The numbers in parentheses are the standard errors of the estimate; see footnote 1.

[a]Asked of nonparent payees of beneficiaries who are aged 18 and older and able to communicate.
[b]Asked of beneficiaries aged 18 and older with nonparent payees.
[c]Asked of payees of beneficiaries who are aged 7 and older and able to communicate.
[d]Asked of all beneficiaries.

SOURCE: Data from the national survey of representative payees and beneficiaries conducted for the National Academies Committee on Social Security Representative Payees (2006).

serving beneficiaries aged 7 and older and who were able to communicate indicated that they had discussed medical, clothing, and food and dietary needs with them. However, among nonparent payees of beneficiaries aged 18 and older, less than a majority, 44.0 percent (2.0), reported having discussed housing needs with them during the past year.

The responses for beneficiaries were similar to those for the representative payees. A majority reported having spoken with their payee about medical, clothing, and food and dietary needs. However, the percentage saying "yes" was slightly lower for them than for the representative payees. While less than a majority of beneficiaries reported having discussed housing needs, the percentage was slightly higher than that reported by representative payees.

Most payees also reported consulting with other people about their beneficiaries' needs at least once a week, 44.2 percent (2.1), or once a month, 22.6 percent (1.4). Most commonly, payees reported discussing beneficiary needs with the beneficiaries' medical or mental health care provider, 68.9 percent (1.7), a relative, 74.2 percent (1.5), another service provider, 37.0 percent (2.1), a friend or neighbor of the beneficiary, 34.2 percent (1.6), or the SSA, 21.3 percent (1.4). A small number of payees reported that they never spoke with others about their beneficiary's needs, 19.4 percent (1.7).

Discussions of Beneficiary Savings

The SSA requires that a payee place any unspent benefit funds in a savings account for the beneficiary. Overall, payees reported that 34.7 percent (1.6) of all beneficiaries had general savings accounts. Of these, slightly less than half of the payees, 48.6 percent (2.3), reported having spoken with their beneficiaries during the past year about saving some of their Social Security benefit money.

Beneficiaries were also asked whether or not they had discussed with their payee placing some of their Social Security benefit dollars into savings accounts. More than half, 57.3 percent (1.7) reported that they had *not* talked with their payees about savings during the past year. Most beneficiaries also reported *not* having any money saved for their future, 58.8 percent (2.1). Among those who did report some savings, the funds came largely from Social Security benefits, 70.6 percent (2.7), money given by family and friends, 45.0 percent (4.2), and from current or past employment, 27.3 percent (3.3). A majority of beneficiaries—57.4 percent (3.7)—identified their payees as the person who usually put aside some of their SSA benefit payments for savings, while almost one-third—30.6 percent (3.1)—reported that they and their payees together put aside some of the benefit payments for savings.

Quality of Communications

A large majority of representative payees reported that their beneficiaries are able to verbally express their needs very well, 59.9 percent (1.4), well, 13.9 percent (1.0), or "okay," 15.1 percent (0.8). For the rest, payees reported that their beneficiaries were able to express their needs verbally "not well," 7.1 percent (0.5), or "not at all," 4.1 percent (0.5).

Among beneficiaries deemed unable to express needs verbally at all or not well, payees also reported that their beneficiaries were able to express their needs *nonverbally* not well, 31.6 percent (3.9), or not at all, 24.1 percent (2.5). The most common explanation for beneficiaries' inabilities to express their needs was a cognitive or intellectual condition, 87.4 percent (3.4). Physical and psychiatric barriers to communication were cited by payees as explanations by 59.7 percent (3.1) and 53.5 percent (5.3), respectively.

Overwhelmingly, payees were either very satisfied, 83.0 percent (1.3), or somewhat satisfied, 14.1 percent (1.2), with their ability to understand their beneficiaries' needs. Only 0.9 percent (0.2) and 0.2 percent (0.1) were dissatisfied or very dissatisfied, respectively. Of these, most, 93.5 percent (2.8) reported that the dissatisfaction was due to the beneficiary's disability. Other reasons for payees' dissatisfaction with understanding beneficiary needs included a language barrier, 20.2 percent (11.1), and generational or other cultural differences, 17.3 percent (8.8).

REPRESENTATIVE PAYEE DUTIES AND RESPONSIBILITIES

It is essential that both payees and their beneficiaries have a clear understanding of the duties and responsibilities with which payees are charged. In this section we examine perceptions of these duties and responsibilities, as well as self-reports of payee performance.

Perceptions of Duties and Performance

Training and support of representative payees are crucial for smooth and effective operation of the payee program. When a payee is recruited, SSA provides a brochure about representative payee duties and responsibilities. This information is also found in the application (www.ssa.gov/online/ ssa-11.pdf [June 2007]), along with a description of the selection process and available training and support for payees. SSA also maintains a website with support materials for representative payees (www.ssa.gov/payee/index. htm [June 2007]).

The SSA specifies—in the brochure and elsewhere—nine responsibilities for representative payees:

1. To ensure that a beneficiary has adequate clothing.
2. To ensure that a beneficiary has adequate housing.
3. To ensure that a beneficiary has adequate food.
4. To ensure that a beneficiary has adequate medical care.
5. To put into a savings account benefit funds not needed for current needs.
6. To file an annual accounting form with SSA.
7. To inform SSA of any changes or events that affect a beneficiary's eligibility.
8. To inform SSA of any changes or events that affect a payee's ability to serve.
9. To return to SSA any payments to which a beneficiary is not entitled.

The vast majority of payees recalled that SSA had informed them of their responsibilities, 94.3 percent (0.6), and most payees have a reasonably complete understanding of their duties. Well over one-half could correctly identify all nine responsibilities, 62.7 percent (2.4); most other payees could correctly identify all but one, 29.5 percent (2.5), or two duties, 4.9 percent

(0.6); and only a relatively small proportion of payees were unable to correctly identify three or more of their responsibilities, 2.9 percent (0.4).

Table 3-2 presents the data on payees' and beneficiaries' knowledge of their responsibilities. As the table shows, knowledge of the complete list of duties is very high among payees with one exception: the requirement to place unspent funds for a beneficiary in a savings account. About one in four payees were unaware of this requirement, or 27.6 percent (2.3). Of course, there may be no funds available for saving, particularly for institutionalized beneficiaries. Only an estimated 35.0 percent (1.7) of beneficiaries have conserved funds for future needs; the amount of such savings was not covered by the committee's survey. Beneficiaries have a level of understanding about payee's responsibilities that is similar to that of the payees themselves. Slightly more than one-half of all beneficiaries, 53.8 percent (3.1), were able to correctly identify all eight of the payee's responsibilities about which they were asked, while over one-third missed one item, 34.7 percent (2.7), and 7.6 percent (1.4) missed two items; only 3.9 percent (0.8) missed three or more of the items. During the committee's

TABLE 3-2 Payee and Beneficiary Knowledge of Payees Responsibilities (in percentage)

Payee Responsibility	Payees Saying Yes	Beneficiaries Saying Yes
Adequate clothing for beneficiary	98.8 (0.3)	97.1 (0.7)
Adequate housing for beneficiary	98.9 (0.3)	99.3 (0.2)
Adequate food for beneficiary	99.4 (0.1)	99.0 (0.3)
Adequate medical care, including mental health care, for beneficiary	98.4 (0.4)	98.3 (0.5)
Putting into a savings account any money left over after paying for current needs	72.5 (2.3)	72.8 (3.1)
Filing an accounting form with SSA every year	92.9 (0.7)	94.1 (1.5)
Telling SSA of any changes or events that affect beneficiary's eligibility for benefits or amount of benefit payments	97.1 (0.4)	95.7 (0.9)
Telling SSA of any changes or events that affect one's ability to serve as a representative payee	97.6 (0.3)	NA
Returning to SSA any payments to which a beneficiary is not entitled	97.7 (0.4)	95.7 (0.7)

NOTES: The nine responsibilities in this table are those listed in the SSA brochure. The numbers in parentheses are the standard errors of the estimate; see footnote 1. NA = not applicable.

SOURCE: Data from the national survey of representative payees and beneficiaries conducted for the National Academies Committee on Social Security Representative Payees (2006).

site visits, SSA staff remarked that one hallmark of a good payee was his or her understanding of the responsibility to report changes in a beneficiary's status to SSA in a timely manner. The SSA lists 11 reportable events:

1. beneficiary moves;
2. beneficiary marries;
3. beneficiary starts working;
4. beneficiary stops working;
5. beneficiary dies;
6. beneficiary's disability condition significantly improves;
7. beneficiary's immigration or citizenship status changes;
8. beneficiary is confined in a correctional institution;
9. beneficiary no longer needs a representatives payee;
10. change in the custodial status of a minor (including adoption); and
11. death or divorce of the parents of a minor beneficiary.

Indeed, most representative payees have a good understanding of when to report changes that affect their own eligibility to serve or their beneficiary's eligibility to receive payments. More than 96 percent of payees knew that each of the 11 events is reportable to SSA. A large majority correctly identified all 11 events necessary to report to SSA, 86.0 percent (1.3), or correctly cited all but one, 3.4 percent (0.7), or two, 3.4 percent (0.7), conditions. Only a very small fraction of payees seems unaware of the full set of events that must be reported to SSA, 2.3 percent (0.5).

Although the high rates of payee and beneficiary knowledge about payees' responsibilities are impressive, it should be noted that they may in part reflect one or more systematic response biases that can influence respondent answers. Those biases include response acquiescence, which is the tendency to agree with survey questions regardless of their content, and social desirability bias, which is the tendency to represent oneself to others (including survey interviewers) in a positive manner. We believe, for example, that acquiescence is evident in the analyses of some questions about fictitious payee responsibilities, such as ensuring that beneficiaries have an interesting hobby and helping take care of a beneficiary's pet. These items were included in the survey in order to assess the degree to which respondents were able to successfully differentiate between duties that are and are not specifically assigned to them by SSA. Clear majorities of both payees and beneficiaries incorrectly identified each of these as payee responsibilities when asked about them along with the actual required responsibilities: This pattern suggests some respondents simply agreed with each of these questions without carefully thinking about their answers. Consequently, the survey data may overestimate payee and beneficiary knowledge of payee responsibilities.

A beneficiary's eligibility requirements are more complicated if he or she also receives Supplemental Security Income (SSI) payments. To assess familiarity with the additional reporting requirements, payees who represent beneficiaries who receive SSI benefits were asked if, to their knowledge, SSA expected them to report when a beneficiary's monthly resources exceed $2,000 (or $3,000 for a couple); when a married beneficiary separates from a spouse or reunites after a separation; when someone moves in or out of a beneficiary's household; when the beneficiary moves to or from a hospital, nursing home, or other institution; and when a beneficiary plans to leave the United States for 30 or more consecutive days. At least 95 percent of all payees correctly answered each of these questions.

COMMUNICATION WITH SSA

Representative Payee and Beneficiary Queries

SSA strongly encourages payees to contact them directly with questions or concerns regarding any aspect of their service or reporting requirements. During the past year, though, few payees, 6.7 percent (0.6), reported seeking help from SSA. Of those who did, most, 74.6 percent (3.4), indicated having done so only once or twice. The most common reported reasons for contacting SSA were to clarify the beneficiary's benefit amount, 46.9 percent (5.8), to understand the payee's responsibilities, 40.2 percent (4.1), and to request permission to allow the beneficiary to manage his or her own SSA benefit payments, 8.7 percent (2.7).

Most payees reported being very satisfied, 43.9 percent (5.2), or somewhat satisfied, 27.6 percent (3.7), with the help they received from SSA. More than one in five, however, indicated being somewhat, 9.3 percent (2.6), or very, 14.5 percent (3.5), dissatisfied with the help they had received.

Some payees, 8.6 percent (0.8) had reported events to SSA during the past year that might have affected the amount of benefit payments made to the beneficiary. Of those who had made a report, in most cases, 64.4 percent (4.7), SSA corrected the benefit amount.

Among beneficiaries, the vast majority, 95.5 percent (0.9), had not sought help from SSA during the past year. When beneficiaries did seek help, it was for a variety of reasons:

- To determine what the payee was supposed to do, 30.8 percent (7.2).
- To understand how to manage benefit funds without a representative payee, 30.7 percent (9.2).

- To learn why a payee was appointed, 25.6 percent (7.8).
- To find out what to do if the payee was misusing the beneficiary's funds, 15.0 percent (6.8).
- To ascertain how their payee was chosen, 18.2 percent (7.5).
- To inquire as to why the payee did not give the beneficiary more money, 13.7 percent (5.0).

Most beneficiaries were either very satisfied, 36.4 percent (6.5), or satisfied, 26.7 percent (7.5), with assistance that Social Security provided when contacted. However, a full 20 percent (7.5) of beneficiaries expressed dissatisfaction with the assistance provided.

SSA Follow-Up

During the committee's site visits, some SSA staff reported having limited resources for following up on questions and problems identified by beneficiaries. Some SSA staff reported that, if a payee receives an SSA benefit check and then disappears, funds for the beneficiary are unavailable for that month. In some instances, SSA staff may be able to assist beneficiaries in taking legal action against payee misuse. However, SSA staff reported not having adequate resources to assess whether a new payee is truly more "suitable" than the old one and often not having resources to double-check the information that potential representative payees provide. Some staff said that if a suitable replacement payee cannot be located, it is better for a beneficiary to have no payee rather than an unsuitable one. Staff also suggested that lack of proper communication among SSA offices may enable a payee who may be unsuitable to "shop" offices in order to become or remain a payee even when one office may have information that disqualifies that person's selection or continuation.

SSA Internet Support for Representative Payees

Although SSA maintains a website that provides information for payees (www.ssa.gov/payee/index.htm), it does not appear to be heavily used. Most payees report Internet access, but only 12.1 percent (1.1) of payees have accessed the SSA website for information on their duties. Low rates of website usage do not appear to be due to lack of access to the Internet, as nearly two-thirds of payees, 61.3 percent (1.4), have access.

For finding information on the website, the majority of payees who used it found it very easy, 47.7 percent (4.6), or somewhat easy, 11.2 percent (2.4). For understanding information on the website, payees rated it as very easy, 51.8 percent (3.5), or somewhat easy, 31.1 percent (3.2). Still, 61,890 (15,738) found it somewhat difficult and 12,367 (4,647) found

TABLE 3-3 Representative Payee and Beneficiary Reports of Joint Responsibility for Paying for Beneficiary Needs (in percentage)

Beneficiary Need	Reported as Joint Responsibility	
	Payees	Beneficiaries
Housing	11.5 (1.4)	16.1 (2.0)
Food and dietary needs	18.2 (1.7)	20.6 (1.9)
Clothing	41.8 (1.7)	35.2 (2.9)
Medical needs	7.4 (0.8)	8.6 (2.2)

NOTE: The numbers in parentheses are the standard errors of the estimate; see footnote 1.

SOURCE: Data from the national survey of representative payees and beneficiaries conducted for the National Academies Committee on Social Security Representative Payees (2006).

it very difficult to locate information on the website, and 27,702 (9,481) found it somewhat difficult and 8,045 (3,700) found it very difficult to understand the information provided.[2]

HANDLING BENEFICIARIES' FUNDS

Beneficiary Needs

Many beneficiaries depend a great deal on their monthly Social Security income. When asked who took responsibility for paying for essential beneficiary needs, most payees indicated that they paid for their beneficiary's housing, 80.6 percent (1.4), food, 69.5 percent (2.0), and medical bills, 73.0 percent (1.7). However, less than a majority, 44.2 percent (1.6), reported being responsible for paying for their beneficiary's clothing, which is apparently viewed as a joint responsibility. Table 3-3 summarizes payee and beneficiary views on shared responsibility for four basic beneficiary needs.

About 80 percent (1.4) of payees pay a beneficiary's housing bills, 67.1 percent (2.0) pay for food, 44.1 percent (1.6) pay for clothing, and 60.2 percent (1.8) pay medical bills. When someone other than the payee took care of paying some of a beneficiary's bills, the most common explanation, mentioned by 35.4 percent (2.5) of payees who did not directly make these payments, was that the beneficiary sometimes functioned well enough to manage his or her own benefit payments.

Estimates based on our survey of beneficiaries 18 years of age and older suggest that the problem of incorrectly appointing a payee is a relatively minor one. The survey showed that beneficiaries and payees agreed that the beneficiary could solely manage his or her payments in only 4.4 percent

---

[2]The raw numbers are weighted estimates.

(0.7) of cases (see Table 6-1 in Chapter 6). We note that it is a violation of SSA rules not to report that a beneficiary no longer needs a payee. In other cases, the beneficiaries lived in a facility that took care of meeting their needs. In these situations, payees signed over the Social Security benefit payments directly to the facility, an arrangement cited by 6.4 percent (0.9) of payees not directly paying bills on behalf of their beneficiaries. In another and slightly larger proportion of cases, in which the payee was not making these payments, 8.8 percent (1.2), someone other than the payee was able to meet the beneficiary's needs. In those instances, payees signed over the Social Security benefit payments to the other person, which is in violation of SSA rules.

Beneficiaries were also asked who they believed usually took care of paying for their needs. As did the payees, the majority of beneficiaries indicated that their payee was primarily responsible for housing, 68.5 percent (2.1), food and dietary needs, 55.2 percent (2.0), and medical needs, 59.6 percent (2.2). Also consistent with the payees' reports, less than one-half of all beneficiaries said that the payee was responsible for purchasing their clothing, 35.2 percent (2.9). Overall, beneficiary reports are largely consistent with payee reports of joint responsibility (Table 3-3). Payees were also asked to estimate how much of their beneficiaries' Social Security benefit payments were used each month to pay for various necessities; average amounts estimated are shown in Table 3-4. Not surprisingly, expenditures for food and shelter represented more than half of expenses reported. Savings represented the smallest category of expenditures reported.

TABLE 3-4 Representative Payee Estimates of Social Security Benefit Expenditures (in dollars)

Expenditure Item	Mean Expenditure
Food	$132.07 (5.1)
Shelter	173.98 (6.5)
Food and shelter (when reported in combination)	324.02 (17.7)
Clothing	82.35 (3.5)
Education and training	38.65 (3.0)
Medical, dental, and mental health expenses	35.71 (3.1)
Recreation and personal items	58.41 (2.6)
Clothing, education, medical and dental expenses, and recreation and personal items (when reported in combination)	236.98 (20.8)
Savings	28.81 (3.6)

NOTE: The numbers in parentheses are the standard errors of the estimate; see footnote 1.

SOURCE: Data from the national survey of representative payees and beneficiaries conducted for the National Academies Committee on Social Security Representative Payees (2006).

Pocket Money

Most payees give their beneficiaries pocket money from their SSA benefits to spend weekly, 45.7 percent (1.3), or monthly, 34.6 (1.6), while some beneficiaries never receive pocket money, 15.4 percent (1.4). Of those provided with pocket money, the average monthly amount received was $96.27 ($10.30).

Consistent with what payees reported, close to one-half, 46.6 percent (1.8), of beneficiaries said their payees gave them pocket money from their Social Security funds once a week, while 36.2 percent (1.5) received pocket money once a month. Beneficiaries, however, reported a higher average monthly allocation of pocket money, $118.97 ($6.26).

Commingling SSA and Other Funds

Most payees, 81.4 percent (1.3), indicated that their beneficiaries did not have sources of income other than Social Security benefits. Most, 75.5 percent (1.3), also reported that none of their beneficiaries' benefit payments from Social Security were saved for future use. When Social Security benefits are saved, typically the payee performs this task, 84.5 percent (1.9). Payees indicated that most of their beneficiaries, 65.4 percent (1.6), did not have a general savings account and that few, 27.8 percent (1.2), ever had their Social Security benefit payments combined with any other person's funds in an account.

Payees whose beneficiaries received SSI benefits were also asked what they did to make sure that savings and other resources did not reach the maximum SSI limit above which beneficiaries are no longer eligible for Social Security benefit payments. The most common strategy mentioned was checking beneficiary resources on a regular basis, reported by 66.1 percent (3.2) of payees. Keeping a budget for the beneficiary was also a common approach, 58.4 percent (2.3). More than one-third of all payees also reported asking their beneficiaries about their resources on a regular basis, 32.2 percent (1.8), and spending some of the money in the account used for savings on things the beneficiary needs in order to maintain the account within allowable limits, 38.2 percent (3.9).

Representative Payee and Beneficiary Agreement on Funds

Most payees, 86.9 percent (1.0), reported that they never had disagreements with their beneficiaries about how Social Security benefit payments were spent. One or two disagreements were reported by 8.1 percent (0.7) of payees, with smaller proportions reporting three to five disagreements, 2.3 percent (0.5), or more than five disagreements, 2.7 percent (0.4). Disagreements tended to be over what benefits were spent on, 59.2 percent (3.7), how much of the benefits were spent, 44.2 percent (3.7), who decided how benefits were to be spent, 48.8 percent (4.3), and about a beneficiary's personal habits, such as smoking, 27.7 percent (3.4).

Similarly, most beneficiaries, 87.6 percent (1.3), reported never having disagreements with their payees regarding how SSA benefits payments were spent. Disagreements once or twice were reported by 8.0 percent (1.2), three to five disagreements were reported by 2.0 percent (0.4), and more than five disagreements were reported by 2.4 percent (0.6). The most common reason for disagreements concerned expenditures for SSA benefits, 65.8 percent (5.0), followed by how much of the benefit should be spent, 54.3 percent (4.8), and who decides how the SSA benefits should be spent, 52.3 percent (4.0). Nearly one-quarter of all beneficiaries, 23.6 percent (4.9), reported having disagreements with payees regarding personal habits, such as smoking.

When viewed together, payee and beneficiary responses are virtually the same (within 1 percent) on the number of disagreements about spending. Beneficiaries reported somewhat more disagreement with their payees than did the payees with their beneficiaries, with beneficiaries having slightly larger discrepancies concerning expenditures of benefits, how much of the benefit should be spent, and who decides.

Representative Payee Record Keeping

About two-thirds of all payees, 68.0 percent (2.1), reported keeping records of how beneficiaries' Social Security funds were spent. Of those keeping records, paper records, such as notebooks or checkbooks, were the most common methods, 76.8 percent (1.6). Less than one in ten payees reported keeping records on a computer, 8.7 percent (0.8).

More beneficiaries believed that their payees kept records of how their Social Security funds were spent, 79.0 percent (1.8), than the payees reported. When records were thought to be kept, beneficiaries also believed that payees were using a notebook, checkbook, or other paper record, 84.8 percent (2.9), or a computer, for 15.5 percent (1.6), which is more than their payees reported.

MEETING BENEFICIARY NEEDS

Ultimately, the purpose of the SSA Representative Payee Program is to ensure that beneficiaries' needs for basic necessities are met. In this section we present the survey findings from payee and beneficiary reports on the degree to which beneficiaries experienced any unmet basic needs during the previous year.

We report, first, however, on payee and beneficiary responses to questions about overall satisfaction with payee performance. Satisfaction with payee performance, on the part of both the payees and beneficiaries, is important not only for the welfare of the beneficiaries, but also for payees' continued willingness to serve. Overall, payees were either very satisfied, 82.1 percent (1.2), or somewhat satisfied, 14.4 percent (1.1), with their ability to help their beneficiaries. Similarly, the vast majority of beneficiaries were also either very satisfied, 86.5 percent (1.5), or somewhat satisfied, 9.0 percent (1.2), with their payees. Only small numbers of beneficiaries were somewhat dissatisfied, 1.1 percent (0.3), or very dissatisfied, 1.2 percent (0.6), with their payees.

Perceptions of Unmet Beneficiary Needs

Table 3-5 presents survey results of payee and beneficiary reports of unmet needs. Although less than 3 percent of payees reported a specific unmet need, 6.0 percent (0.6) of payees reported having a beneficiary with at least one unmet need. Similarly, less than 3 percent of beneficiaries reported any specific unmet need, but somewhat more of them reported having at least one unmet need during the previous year, 7.2 percent (1.0). Consistently, however, beneficiaries were more likely to report unmet needs than were their payees. In fact, although the number of these reports were small overall, beneficiaries were twice as likely to report having been without utilities

TABLE 3-5 Representative Payee and Beneficiary Reports of Unmet Beneficiary Needs During the Past (in percentage)

Unmet Need	Payees Saying Yes	Beneficiaries Saying Yes	Beneficiary/ Payee Ratio
Place to live	0.7 (0.2)	1.0 (0.4)	1.4
Utilities, including heat, water, and electricity	1.0 (0.2)	2.2 (0.5)	2.2
Food (been hungry)	0.8 (0.2)	2.4 (0.6)	3.0
Clothes	0.5 (0.2)	2.0 (0.5)	4.0
Seeing a doctor, dentist, therapist, eye doctor, or other medical professional	2.8 (0.4)	2.8 (0.6)	1.0
Medication	1.8 (0.3)	2.8 (0.6)	1.6
Any need	6.0 (0.6)	7.2 (1.0)	1.2

NOTE: The numbers in parentheses are the standard errors of the estimate; see footnote 1.

SOURCE: Data from the national survey of representative payees and beneficiaries conducted for the National Academies Committee on Social Security Representative Payees (2006).

or having gone hungry at some point during the past year, and three times as likely to report having needed (and not received) clothing as their payee counterparts.

Payees were also asked if they believed that their beneficiaries had all, most, some, or none of their needs met across the same domains during the past year. As shown in Table 3-6, more than 90 percent of payees reported that they believed all their beneficiaries' needs were met in four of the six domains: housing, food, clothing, and medical needs. However, about one in five payees, 19.7 percent (1.1), reported that none of the education or training needs of their beneficiary were being met.

As shown in Table 3-7, beneficiaries' responses about unmet needs par-

TABLE 3-6 Payee Perceptions of Beneficiary Needs Met During the Past Year (in percentage)

Need	All of the Time	Most of the Time	Some of the Time	None of the Time
Housing, including utilities	94.4 (1.0)	3.3 (0.6)	2.0 (0.4)	0.3 (0.1)
Food	95.4 (1.1)	2.7 (0.7)	1.9 (0.4)	0.1 (0.1)
Clothing	92.0 (1.1)	4.6 (0.7)	3.2 (0.5)	0.2 (0.1)
Medical	93.1 (1.1)	3.0 (0.5)	3.0 (0.5)	0.8 (0.3)
Recreation	81.9 (1.1)	7.1 (0.6)	8.6 (0.7)	2.4 (0.4)
Education or training	74.6 (0.9)	3.2 (0.6)	4.2 (0.4)	18.0 (0.9)

NOTE: The numbers in parentheses are the standard errors of the estimate; see footnote 1.

SOURCE: Data from the national survey of representative payees and beneficiaries conducted for the National Academies Committee on Social Security Representative Payees (2006).

TABLE 3-7 Beneficiary Perceptions of Their Needs Met During the Past Year (in percentage)

Need	All of the Time	Most of the Time	Some of the Time	None of the Time
Housing, including utilities	94.1 (1.2)	3.5 (0.8)	2.0 (0.6)	0.4 (0.2)
Food	93.4 (1.2)	4.2 (0.9)	2.2 (0.5)	0.2 (0.2)
Clothing	92.6 (1.1)	3.9 (0.7)	2.8 (0.5)	0.8 (0.3)
Medical	93.5 (1.0)	3.9 (0.8)	1.9 (0.4)	0.7 (0.3)
Recreation	83.2 (1.6)	5.9 (0.8)	7.1 (0.9)	3.8 (0.5)
Education or training	74.9 (1.3)	3.9 (0.7)	2.8 (0.6)	18.3 (1.2)

NOTE: The numbers in parentheses are the standard errors of the estimate; see footnote 1.

SOURCE: Data from the national survey of representative payees and beneficiaries conducted for the National Academies Committee on Social Security Representative Payees (2006).

alleled those of payees. More than 90 percent of beneficiaries reported that all of their needs for housing, food, clothing, and medical care had been met. Like the payees, nearly one in five beneficiaries, 18.3 percent (1.2), reported that none of their education or training needs had been met.

During the committee's site visits, some SSA staff stressed that beneficiaries are reluctant to report issues of unmet need, either resulting from a payee's failure to address known problems or a payee's failure to exercise discretion regarding beneficiary funds. Some staff noted limited resources for the office to follow up on problems brought to their attention.

Agreement Between Beneficiary and Representative Payee Reports

The survey findings reported in this chapter show close agreement between payees and beneficiaries. However, these comparisons are at the aggregate level and do not necessarily indicate the level of agreement between pairs (dyads) of beneficiaries and payees. To assess the consistency of reporting between individual beneficiary and payee dyads, we examined the percent agreement[3] between payee and beneficiary reports. Table 3-8 presents the agreement statistics; responses that are payee requirements by SSA are indicated by the symbol @. The data show that more than 95 percent of all payee-beneficiary dyads provided consistent responses on six questions about unmet needs. The percent agreement in responding to communication and financial practices questions was not as high, but still indicated that close to two-thirds or more of all payee-beneficiary dyads provided identical reports to those items.

CONCLUSIONS AND RECOMMENDATIONS

Communication

Most payees indicated communicating with beneficiaries at least once a week, most often through face-to-face meetings, followed by telephone calls. The majority of payees consulted with others about their beneficiaries' needs weekly, most often with medical or mental health professionals, and also with relatives. For beneficiaries with savings accounts, one-half of the payees had not spoken with their beneficiary about saving money. For those payees having discussed savings in the past year, slightly over one-half discussed it once every few months.

Payees reported that nearly half of beneficiaries express their needs very well. One-third of payees thought that their beneficiaries were un-

---

[3]The percent agreement measure is simply the proportion of cases in which both the payee and the beneficiary gave the same answer.

TABLE 3-8 Agreement Between Reports of Representative Payee and Beneficiary Dyads (in percentage)

Domain	Agreement
Communication Between Payee and Beneficiary	
Saving some Social Security benefit money	67.6 (3.5)
@Housing needs	64.3 (2.2)
@Food and diet needs	64.3 (2.2)
@Clothing needs	67.7 (1.9)
@Medical needs	74.7 (1.7)
Financial Information and Practices	
Beneficiary has other sources of income	86.4 (1.2)
Beneficiary has a general savings account	70.2 (1.8)
Payee keeps records of how SS benefits are spent	68.9 (2.0)
Unmet Needs During the Past Year	
Housing	98.6 (0.4)
Utilities	97.3 (0.6)
Hungry (been hungry)	97.3 (0.6)
Clothes	98.0 (0.5)
Seeing a doctor, dentist, therapist, eye doctor, or other medical professional	95.1 (0.6)
Medication	96.5 (0.7)

NOTES: The numbers in parentheses are the standard errors of the estimate; see footnote 1. @ = SSA requirement.

SOURCE: Data from the national survey of representative payees and beneficiaries conducted for the National Academies Committee on Social Security Representative Payees (2006).

able to express their needs nonverbally. The most common impediments to beneficiaries' communication were cognitive or intellectual conditions. Most payees were very satisfied with their ability to understand beneficiary needs. For a small percentage, dissatisfaction was mainly attributed to the beneficiary's disability.

Most beneficiaries not living with the payee had weekly contact with them. Very few had no contact with the payee. Most beneficiaries said they were contacted in person, with the second most frequent mode being by telephone.

For the few payees who had questions for SSA about their role, inquiries tended to concern the correct benefit amount for the beneficiary and further clarification of their responsibilities. One in five payees who contacted SSA reported being dissatisfied with the assistance provided. Also, one-fifth of the beneficiaries who contacted SSA were not satisfied with the assistance that SSA staff gave them. Although the number of payees and beneficiaries who seek help is small, the fact that more than one in five who sought help are dissatisfied to some extent is of concern and bears further investigation by SSA.

The website that SSA provides for assistance to payees does not appear to be used very much, in spite of a significant number of payees who report having access to the Internet. The lack of use may be due to a lack of knowledge of the site's existence or to how information on the website is presented. In either case, the website's usability (e.g., user friendliness, readability) should be periodically reviewed and updated by SSA staff, and its existence regularly communicated to representative payees.

Meeting Beneficiary Needs and Saving Funds

Over one-half of beneficiaries had not spoken with their payees about putting some Social Security benefit dollars in savings, and most had no money saved for their future, although one-quarter had saved some amount of benefit payments. However, many payees were unaware that they could or should place unused funds in a savings account. Because social security benefits are often the only source of income for most beneficiaries, it is important that payees use funds for the needs of beneficiaries and conserve them when at all possible.

RECOMMENDATION 3.1 The Social Security Administration should strengthen its efforts to encourage representative payees to save money for beneficiaries and to enforce the requirement that the saved money is put in a specified savings account.

One additional area for scrutiny regarding beneficiary needs is that of their education or training needs, an SSA-required payee responsibility that was reported not met by one-fifth of payees and beneficiaries. Payee's attention to beneficiary education and training and to savings (when possible) should be a priority, one that SSA may want to strongly emphasize to representative payees.

Information from the committee's site visits is instructive regarding payees' meeting beneficiaries needs. Some SSA staff said that beneficiaries were reticent to come forward to report problems due to a fear of potential retribution by the payee. Moreover, if the beneficiary is a person with a communication limitation, his or her ability to report to SSA is further compromised.

Representative Payee Duties and Performance

Most payees understood that they should provide beneficiaries with adequate housing, food, and clothing, but many did not appear to understand that they were to deposit any leftover payments into a specified account (see above).

Over one-third of payees said that they usually give the beneficiary pocket money, though a few never do. For beneficiaries receiving SSI payments, well over one-half of payees checked beneficiaries' resources on a regular basis and maintained a budget for them. Most payees and beneficiaries reported having no arguments regarding how Social Security payments were spent. For the few beneficiaries with arguments, more than one-half were concerned about how the money would be spent. Again, one must be cautious in interpreting the level of agreement between payees and beneficiaries because, as stressed by SSA staff during our site visits, beneficiaries may be reticent to report disagreements with their payees.

Most payees took responsibility for paying for essential beneficiary needs (e.g., housing, food, medical bills). For many payees and beneficiaries, this was a shared responsibility. In many of the instances in which the payee did not directly make payments, the beneficiaries did so or the facilities in which they lived did so.

The committee concludes on the basis of its survey and site visits that payees are generally performing their duties in accordance with SSA standards. According to the more in-depth information from dyads of payees and their beneficiaries, they maintain reasonably frequent contact and discuss basic needs. The dyads are also in general agreement as to who pays for the beneficiary's basic needs, whether or not the beneficiary receives pocket money, whether or not the payee maintains records of expenditures, and whether or not beneficiaries have experienced unmet needs in the past year. Payees and beneficiaries appear to have high levels of understanding of the basic duties and responsibilities of payees, with the exception of the requirement to put into specified savings accounts funds that are not required for current needs. Both payees and beneficiaries report high levels of satisfaction with payee performance.

4

Defining and Discovering Misuse

This chapter addresses part of the third of the four specific items in the committee's charge: "identify the types of representative payees that have the highest risk of misuse of benefits." In order to address this charge, we first had to define misuse and understand how it differs from program violations and improper use of Social Security benefits; this information is covered in the first section of this chapter. The next section looks at the current methods of the Social Security Administration (SSA) to identify and monitor misuse. Because little if any misuse is currently detected or reported, the committee carried out two in-depth examinations of payee records, one on lump-sum payments and the accounting form and a separate in-depth study of misuse. The results of this work are covered in the final sections of this chapter.

DEFINING MISUSE

Formal Definitions

For many years, misuse was not formally defined by SSA or by the Social Security Act. Rather, there were guidelines on the correct use of benefits by a representative payee. It was not until March 2004 that Congress enacted the Social Security Protection Act (P.L. 108-203) that defined misuse by a representative payee for either or both Old Age, Survivors, and Disability Insurance (OASDI) benefits, generally referred to as Social Security, and Supplemental Security Income (SSI):

> Misuse occurs in any case in which the representative payee receives payment under this title for the use and benefit of another person and converts such payment, or any part thereof, to a use other than for the use and benefit of such other person.

SSA also defines misuse in its Program Operations Manual System (POMS), which is used by SSA staff to administer all SSA programs. In "Overview of Misuse of Benefits" (GN 00604.001), misuse is defined as the opposite of what is required:

> A representative payee (payee) is required to use the benefits only for the use and benefit of the beneficiary. Misuse occurs when the payee uses the benefit for any other purpose.

In the same POMS document, misuse is again characterized:

> Misuse of benefits refers to the misappropriation of benefits by the payee. Misuse of benefits occurs when the payee neither uses the benefits for the current and foreseeable needs of the beneficiary, nor conserves benefits for the beneficiary.

The document continues to distinguish between misuse and improper use:

> Improper use of benefits is an unwise expenditure of benefits in a manner that is not in the beneficiary's best interest. Because the beneficiary still receives a benefit from the expenditure, this is not a misuse.

The definition of misuse seems simple, but to identify misuse is a complex undertaking. One reason for the complexity is that the failure of a representative payee to perform some required duties (see Chapter 1) may constitute a violation but not misuse. SSA does not give a formal definition of a violation, and there are no penalties for violations.

Distinguishing Between Violations and Misuse

A typical example of the difficulty of distinguishing misuse and violation is the commingling of a beneficiary's funds with those of the representative payee. This is clearly a violation of payees' duties, but it is not a misuse of OASDI or SSI funds unless the payee uses the beneficiary portion of the funds for someone other than the beneficiary.

Another illustration is taken from our survey. For one-half of the representative payees, we planned to interview one of their beneficiaries. In many cases, the beneficiary lived with the representative payee and, if not, the representative payee would tell us how to contact the beneficiary and help the interviewer arrange a time to interview the beneficiary. Surprisingly, however, we found nine cases in which the representative payee had no idea where the beneficiary could be located. Moreover, even after a careful review of all factors in the situation and using extra tracing efforts, the interviewers could not find those beneficiaries. Yet the benefit checks for those beneficiaries were still being issued.

In these cases, we could not determine if the representative payee was saving the funds for the beneficiary or if the funds were being misused. It was clear that the benefits were not being immediately used to help the beneficiary and therefore that at least a violation had occurred. At a minimum, the payee was obligated to notify SSA that the beneficiary had changed his or her address to some unknown address. Using the survey weights (see Appendix A) we estimate that, annually, nearly $54 million in benefits are sent to representative payees who do not know where the beneficiary can be located. It is likely that at least some of these cases represent misuse.

To illustrate further, suppose a representative payee does not report a change, such as the beneficiary's going to jail, that could result in a change to the beneficiary's eligibility. (When a beneficiary is in a penal institution, s/he is not entitled to receive OASDI or SSI benefits.) If the payee does not report this change and continues to accept the funds, there may be misuse. If the funds are spent on people other than the beneficiary, it is misuse; but if the money is saved for the beneficiary, it is an overpayment by the SSA, which is a violation but not misuse.

The distinction between misuse and a violation is sometimes difficult to understand. An overpayment carries the connotation that SSA made a mistake and sent the payee too much money. However, the money was disbursed because the payee did not notify SSA that the beneficiary's circumstances had changed. That is, the representative payee did not fulfill one of the specified duties. However, breaking rules does not necessarily meet the formal definition of misuse. Because identifying misuse is complex and there are few guidelines from SSA on it, claims representatives are more likely to find a "more suitable" payee rather than to label something an overpayment or to report it as suspected misuse.

CURRENT WAYS OF DETECTING MISUSE

Given the size and scope of the Representative Payee Program, as well as the physical or mental health problems of many beneficiaries, a program to detect misuse is critical. Currently, there are three main avenues for detecting misuse: information given to SSA, studies by the Office of the Inspector General (IG), and the annual accounting form.

Information Provided to SSA

Information may be provided to SSA by beneficiaries themselves or a concerned third party. Frequently, SSA will be contacted by a person who believes that she or he should be the representative payee because the person has physical custody of the beneficiary but is not receiving the benefit payments. Often, a dispute has arisen between relatives of a beneficiary about whom the payee is and who should be receiving funds. These cases are handled by the local SSA staff. Sometimes, misuse is found. More often, any irregularity that is found is termed an overpayment or a payee change is made. SSA claims representatives are not trained as social workers and do not feel it is their role to investigate and resolve these issues.

Claims representatives find that many of the problems brought to their attention do not fit the definition of misuse although they are clearly violations of the program. For example, a beneficiary might be ineligible to receive SSI funds because she or he is working, but the benefit check is still going to the payee. This situation is likely to be classified as an overpayment even though it is a payee's responsibility to tell SSA when a beneficiary's status changes. Claims representatives seem to be more comfortable in assigning an overpayment code than a misuse code. They also often told us their beliefs that unless the amount of money involved was above a certain amount (we heard $20,000 most often) the IG would not pursue the case.

Claims representatives frequently noted in the Representative Payee System (RPS) that a more suitable payee was found. However, in many cases, the original payee, who had been replaced, subsequently was reappointed. This often happens in custody cases when the physical custody of a beneficiary switches from one parent to another. However, in some cases, we were told that payees "shop" local SSA offices because some offices do not document misuse effectively or use the RPS for this purpose. Thus, an SSA office may not know the history of a payee who has been replaced or terminated because of suspected or known misuse or violations, which then may continue.

CONCLUSION Relying on beneficiaries or third parties to report misuse to Social Security Administration is not a reliable or efficient primary strategy for detecting misuse.

Office of the Inspector General

The second avenue of discovering misuse is located in the IG.[1] In this program, small simple random samples (about 250 cases) of the payee

---

[1]The SSA IG operates independently of SSA programs and investigates all programs for fraud and irregularities.

population are occasionally selected. The office then audits these payees, reviews their finances, and issues a report. From the various reports issued by the IG, apparently no misuse has ever been found in these samples.

The most recent report, *Nation-wide Review of Individual Representative Payees for the Social Security Administration* (U.S. Social Security Administration, 2005), reviewed 275 payees who served 14 or fewer beneficiaries, a total of 359 beneficiaries). The IG confirmed the existence of all 359 beneficiaries and concluded that the needs of 356 of them were met. For the remaining three cases, no conclusions were drawn. The IG's report also found that 13 of these payees did not follow SSA policies, 8 acted as conduit payees,[2] and 5 failed to report events that could have affected the amount of benefits received or even the beneficiaries' rights to receive benefits.

In addition, the IG found five cases (2 percent) of inaccurate data in the RPS, including different spellings of the same individual's name, different addresses and phone numbers, post office boxes listed as the payee's residence address, and wrong addresses. These discrepancies are far fewer than the more than 20 percent inaccurate addresses that Westat encountered in its survey operations (see Appendix A).

CONCLUSION The methodology used by the Social Security Administration Inspector General does not detect misuse.

The Annual Accounting Form

The third avenue for detecting misuse is through the annual accounting form. As noted in Chapter 1, this form is required to be filled out annually by each payee for each beneficiary (or group of beneficiaries if children). See Appendix E for a copy of the form. The total amount of benefits received during the 12-month accounting period is preprinted on the form supplied by SSA. The payee is asked to show how much was spent in each of a few categories. Though completion of the form is mandatory, it is not always filed in a timely way. The forms are routed to a central processing center where they are reviewed, potential errors are circled, and those with potential errors are sent to an SSA Program Service Center for resolution. In our site visits and other interviews with SSA staff, we were told that since the inception of the accounting process more than 30 years ago, very little misuse appears to have been detected by this method.

CONCLUSION The Social Security Administration does not discover misuse by using the annual accounting form.

---

[2] A conduit payee simply passes the monthly benefit check directly to another person without exercising a payee's duties to manage the funds.

IN-DEPTH STUDIES TO UNDERSTAND POSSIBLE MISUSE

From our site visits, the committee learned that SSA staff believe that considerable misuse is occurring but that most of it is not reported as misuse. Staff said that it requires at least 10 hours to document a typical misuse case and that they received no encouragement or incentives to document it. Thus, the staff tends to deal with misuse by either changing payees or by classifying the cases as overpayments.

In part as a result of this information, the committee initiated studies to learn how SSA characterizes and discovers misuse. One study looked at lump-sum payments and the accounting form. We also undertook an in-depth study of SSA files that documented the circumstances of misuse by representative payees and then looked at the results of our survey to see if misuse could be detected by examining how well the needs of beneficiaries are being met.

Lump-Sum Payments and Annual Accounting

The committee decided to take a closer look at the accounting process and to examine how the accounting process works for beneficiaries who are recipients of large retroactive sums (lump-sum payments). Logic dictates that SSA would monitor more closely representative payees who receive large lump-sum payments on behalf of their beneficiaries. As noted in Chapter 1, a large amount of money to an individual can be involved, especially if there is a delay in determining disability status. However, SSA does not total all lump-sum payments, and we could obtain no official statistics. We used different strategies to summarize the OASDI and SSI payment data and estimated, with very bad administrative data, that more than $700 million per year is paid as lump sums.

The committee looked first at SSI lump-sum payments. During the period from 2000 to 2004, the largest lump sum paid to a payee on behalf of an SSI recipient was $125,000; this amount was received by an organization with 29 beneficiaries. The largest payment to an individual SSI payee was $67,130. However, most payees received an average lump-sum payment of less than $3,000; the highest frequency of payments was in the $1,000 to $1,500 range, which would equal the benefits for a couple of months. (The average monthly federally administered SSI amount was $428.29 in that time period.)

For OASDI beneficiaries, the committee examined the payment streams for beneficiaries with individual representative payees who (1) received more than $100,000 in the 60-month period from January 2000 through December 2004 and (2) received one or more monthly payments for which the amount was at least twice the amount for the previous month.

For OASDI, the largest single lump sum paid to an individual payee was $229,563. For organizational payees, the lump sums ranged from less than $10,000 to more than $100,000. On average, the payees received lump-sum amounts equivalent to 2 or 3 month's worth of benefits.

According to the RPS, none of the payees in our lump-sum study were labeled as misusing funds. Yet, our review of accounting forms for a sample of 100 of these beneficiaries revealed many examples of inappropriate use if not blatant misuse. For, example on the back of an accounting form was the following statement:

> A beneficiary gets $70,687 in a lump-sum payment. The daughter is the payee. She says, "The beneficiary gave his daughter a minimum of $40,000 to pay off her debts and purchase a house. He gave his brother $10,000 to pay off a debt he had. He gave his granddaughter a $5,000 gift for her wedding and he also gave his grandson $2,000 so he could purchase a car."

This is patently a case of misuse.

Because the total annual benefit amount is preprinted on the annual accounting form by SSA, it is all too easy for a payee to merely match the claimed spending totals to that amount. Moreover, from the review of the forms, it appears that one can actually put down a *different* amount of spending and not fail the review as long as the claimed expenses exceed the expected amount. For example, on one approved form, the preprinted SSA benefit amount was $26,102. The payee wrote that $58,596 was spent on food and housing and $1,523 on other expenses.

We also found that SSA does not appear to keep track of money in savings accounts. For example, in one year a payee put $13,000 of a lump-sum payment of $83,067 in savings; the next year the approved annual accounting form showed that the savings amount reported the previous year was $0, not $13,000.

Sometimes, payees appear either unable to control the spending or unable to decide how all the money is spent—even though this task is one of their primary responsibilities. This statement was noted in another annual report we reviewed:

> Beneficiary spends it any way he wants. He gives me $450 per month and he is paying $75 per month on a television. The rest of his check he throws away on whatever. I don't know on what. I took $1,000 out of his account to buy him four brand new tires for his car and also two new suits. (Mother is the payee and the lump-sum amount was $50,128.)

Some payees might report information on the back of the annual accounting form but not on the front. One payee left everything blank on the front of the form and wrote this statement on the back:

I cash my brother's check and give the money to him. He takes care of his living expenses. To my knowledge he has no money in savings. His counselor said he was able to take care of his finances and paperwork has been filled out and sent to Social Security to that effect. (Lump-sum amount was $14,119 and the annual report was approved.)

CONCLUSION As a practical matter, the Social Security Administration does not require representative payees to carefully account for dollars spent in each category on the annual accounting form as long as the total amount spent is approximately the same or greater than the total amount of benefits received.

CONCLUSION The Social Security Administration does not apply special monitoring to payees of beneficiaries who receive lump-sum payments.

CONCLUSION Large lump-sum payments appear to be spent on items or persons that would not be approved by the Social Security Administration and would be considered a misuse of funds.

RECOMMENDATION 4.1 The Social Security Administration should give special scrutiny to representative payees who receive lump-sum payments.

The committee suggests three changes to better monitor lump-sum payments: (1) adding a lump-sum indicator and amount to the RPS, to enable the local field office to monitor how the lump-sum money is spent for a specific beneficiary, (2) revising the POMS to include special, additional accounting protocols whenever lump-sum payments are issued to a beneficiary who has a representative payee, and (3) modifying the annual accounting form to track specifically how lump-sum funds are spent and saved.

In-Depth Study of Misuse

The in-depth study of misuse was undertaken at the urging of SSA. SSA staff actively helped conduct this study, creating folders of materials and reviewing their contents. Detailed results of the study are shown in Appendix D. Though the tables in the appendix provide information about misusers, many of the statistics based on these data do not provide any insight because of the large amount of "not reported" or "not available" information. Even though the SSA staff diligently searched for information about misuse events, when they assembled the folders they found that misuse is not well documented in SSA records.

The lack of documentation of misuse events in the RPS makes it difficult for SSA to learn about the demographic and socioeconomic characteristics of the payees who are labeled as misusers or about the causes of misuse. As a consequence, it is difficult for SSA to make empirically based decisions about possible changes to policies and procedures pertaining to selection, training, and monitoring representative payees.

A surprising discovery from this study was that some payees who have a history of misuse are still active as payees. SSA policy does not prohibit a prior misuser from becoming or remaining a payee (POMS GN 00501.132, Section B.2.d). In fact, many (though not all) of these misusers are mothers of underage beneficiaries, and in such cases, it may be that more appropriate payees are not available (see Appendix D).

RECOMMENDATION 4.2 The Social Security Administration should develop new procedures and policies that prevent the routine reappointment of a representative payee who has been documented as a misuser or a continued violator of Social Security Administration policies and rules

When circumstances dictate retention or reappointment of such payees (e.g., a parent-child situation) the committee suggests that SSA should justify the action in writing in the RPS. SSA should ensure that the performance of these payees receives greater scrutiny than other payees, such as semiannual, in-person interviews to review accounting records.

Using Beneficiaries' Questionnaire Responses to Find Misuse

Because misuse is defined as the use of a beneficiary's Social Security funds for someone other than the beneficiary, some cases of misuse could lead to actual deprivation of beneficiary funds critical for basic needs such as shelter, food, and medical care. Thus, an important question with respect to misuse is whether a reported lack of shelter, food, or medical care for a beneficiary is an indicator of misuse by the payee.

To address this important issue, questions about unmet needs were included on the surveys of representative payees and beneficiaries. Specific questions attempted to determine if there had been specific instances when the beneficiary had not had shelter, food, medical care, or other basic needs met within the last 12 months; others were more subjective, asking if the beneficiary had an unmet need all, most, some, or none of the time.

When the responses to the factual and subjective questions were cross-tabulated, they were, in general, consistent. That is, beneficiaries claiming a

TABLE 4-1 Comparison of Beneficiary Responses: Factual and Subjective Questions About Unmet Medical Needs (in percentage)

Factual Question: Were you ever unable to get medical attention when you thought you needed it?	Subjective Question: Were your medical needs met?		
	None or Some of the Time	Most or All of the Time	Total
No	1.6	98.4	100.0
Yes	29.5	70.5	100.0

SOURCE: Data from the national survey of representative payees and beneficiaries conducted for the National Academies Committee on Social Security Representative Payees (2006).

TABLE 4-2 Comparisons of General and Specific Beneficiary Responses on Needs Met by Specific Need

Need	Some or None of the Time	Actually Went Without	Percent Went Without
Housing	66,689	7,756 (4,432)[a]	11.6 (6.8)
Utilities	67,994	23,596 (8,782)	34.7 (11.0)
Food	67,948	26,799 (11,200)	39.4 (11.8)
Clothing	100,329	35,953 (11,612)	35.8 (9.2)
Medical care	67,501	23,076 (7,480)	34.2 (9.1)
Medicine	67,501	15,786 (6,667)	23.4 (8.5)

[a]See footnote 3.

SOURCE: Data from the national survey of representative payees and beneficiaries conducted for the National Academies Committee on Social Security Representative Payees (2006).

specific instance of an unmet need also stated that same need was met only some or none of the time significantly more than those without a specific instance. Table 4-1 provides an example of the cross-tabulations, focusing on medical care.

Table 4-2 provides a summary of the cross-tabulations. Note that the proportion of those whose housing needs were met only some or none of the time and who actually experienced no place to live within the last 12 months is not significantly different from zero, 11.6 percent (6.8).[3] In most other cases, the survey results indicate that about one-third of those beneficiaries who said some aspect of their needs were met only some or none of

---

[3]For all estimates in this chapter, the number in parentheses following the estimate is the standard error of the estimate. As a general rule, users can approximate a 95-percent confidence interval for the estimate by adding and subtracting two standard errors to the estimate. When two estimates have confidence intervals that overlap, the two estimates are not statistically different at the .05 level of significance.

the time actually experienced a period in the last 12 months when they had unpaid utility bills, went hungry, or were unable to pay for needed medical services or medicine.

These estimates need to be viewed cautiously because they are derived from only a small number of cases. That is, only 37 beneficiaries provided consistent responses to the factual and subjective questions on at least one type of unmet need. These cases were studied in more depth, but no conclusions about misuse were possible, given the availability of only survey data. Due to the criteria for selecting cases for reinterview, none of the representative payees for these 37 beneficiaries were included in the reinterview survey.

CONCLUSION

The committee undertook many tasks in examining misuse and its characteristics—an in-depth study of known misuser cases, an examination of how lump-sum payments are monitored, and an attempt to use survey data to uncover misuse. It was apparent from all of these tasks that a new approach is needed to identify the types of representative payees that have the highest risk of misuse of benefits. In the next chapter we present a methodology to discover misuse and thus, ways to reduce the risk of misuse and better protect beneficiaries.

5

New Approaches to Detect Misuse

This chapter continues the committee's response to the third of four specific items in its charge and addresses the fourth item: "to identify the types of representative payees that have the highest risk of misuse of benefits" and to "find ways to reduce the risk of misuse of benefits and ways to better protect beneficiaries." The first section in this chapter presents a new strategy the committee developed to identify the types of representative payees that have the highest risk of misuse of benefits and a methodology to identify misusers. The second section examines four common violations of Social Security Administration (SSA) policies that would not be classified as misuse by the current definitions but raise questions about whether they mask misuse. The third section examines the annual accounting form and its possible role in better protecting beneficiaries. The fourth section considers the use of bank-account-linked debit cards for use in the Representative Payee Program as a way to better serve and protect beneficiaries. In the last section we consider the role of the program staff structure in terms of detecting misuse and protecting beneficiaries.

A NEW METHOD TO DETECT MISUSE

As detailed in the previous chapter, none of the methods currently used by the SSA to detect misuse appears effective. Based only on identified misusers in SSA administrative records, the incidence of misuse is about 0.01 percent in a population of about 5.4 million representative payees.

Indeed, Fritz Streckewald, Acting Assistant Deputy Commissioner for Disability and Income Security Programs at SSA, said in testimony before the House Ways and Means Subcommittee on Social Security on May 10, 2001, that "misuse of funds occurs in less than one-hundredth of one percent of cases." Moreover, the SSA's Office of the Inspector General (IG) studies of misuse that are based on small, simple random samples of payees have, not surprisingly, never found a case of misuse. The goal in this phase of the project was to determine if the committee could find a method to enable SSA to efficiently select individual representative payees for audit to detect misuse of beneficiary funds.

Determining Possible Indicators of Misuse

The committee began by looking at the available administrative data, analyzing our in-depth study of misusers (see Appendix D), and talking with SSA field and headquarters staff. The committee also reviewed the methodology used by the SSA IG, and, for comparison, the methodology used by the Internal Revenue Service for selecting taxpayers for audit.

The committee then selected 15 variables from the administrative records that seemed to be indicators of possible misuse of beneficiary funds by a representative payee.

1. Payee is a nonrelative.
2. Payee does not live with beneficiary.
3. Payee is under 50 years of age.
4. Payee has sources of income other than employment.
5. Payee receives welfare (TANF[1]).
6. Payee lists self-employment income.
7. Payee receives either OASDI or SSI income or both.
8. Payee is a convicted felon.
9. Payee has served time in prison.
10. Payee's mailing and residence addresses differ.
11. Payee has had three or more address changes in the last two years.
12. Payee does not have a phone number in the administrative records.
13. Payee has been terminated two times or more.
14. Payee serves for four or more beneficiaries.
15. Payee lives in different zip code from beneficiary.

---

[1]Temporary Assistance for Needy Families (TANF) took effect on July 1, 1997. It replaced Aid to Families with Dependent Children and the Job Opportunities and Basic Skills Training programs. It provides assistance and work opportunities to needy families.

Each variable was scored as a 1 (payee had the characteristic) or 0 and summed for all representative payees in the sample.

Unfortunately, the administrative data in the Representative Payee System (RPS) often proved to be wrong, leading to some committee concern about the accuracy of the scores. For many representative payees, several variables were blank or missing: in those cases, a de facto score of 0 was assigned for those variables, thus lowering the probability of scoring high on the scale of 0 to 15.

The committee analyzed the records of more than 5,000 representative payees. Of these payeees, 288 were deemed to have high scores and were selected to be the basis of a reinterview group to be studied in more detail as potential misusers.

Reinterviews: Process

The committee did not have the time or the funds to reinterview all 288 cases. Therefore, the committee identified six states (California, Connecticut, Florida, Ohio, Oregon, and Texas) in six SSA regions that each contained a large number of cases. This step led to the identification of 99 of the 288 cases for reinterviews. (All but three of the payees had been interviewed in the initial survey of representative payees.)

Because the reinterviews would entail asking detailed questions about finances, banking accounts, and expenditures, two-person teams were used for them. One was a person with financial expertise; the other was a senior social scientist.

The process began with a certified letter signed by an SSA official notifying the representative payee of the location and time of the reinterview. Because the letters were certified, the payees had to sign for them. The letter stated that the reinterview was mandatory, identified the location and time of the reinterview, pledged confidentiality, and listed the documents that the payee should bring. Because of SSA office procedures, all reinterviews had to be scheduled during normal working hours.

In the next step, shortly after the letters were sent, staff of the survey firm working for the committee (see Chapter 1), Westat, made telephone calls to the recipients to ensure that the letters had been received and confirming the appointments. During these follow-up telephone calls, the payees were reminded of what documents were required. The final preinterview step was a reminder call from Westat staff to the representative payees a few days before the scheduled reinterviews.

Before the reinterviews, the teams reviewed all the financial data available from SSA and noted any questions to be raised with the payees. After the reinterviews, the teams wrote reports that were sent to Westat for review within 36 hours. These reports noted whether or not the team had categorized the case as misuse, uncertain or potential misuse, or no misuse. The reports were then forwarded to the committee for its review. The committee review involved detailed examinations and discussions, in conference calls and, in some cases, again at committee meetings.

Reinterviews: Results

Of the 99 cases selected for the reinterview process, 76 could be analyzed: 4 payees could not be found (although two of them had been interviewed in the survey 3 months earlier), and 19 did not show up for their scheduled reinterviews.

Of the 76 cases, the committee determined that 16 cases involved misuse, 17 showed uncertain or potential misuse, and 43 did not involve misuse. In reviewing the data on those 76 cases, we looked separately at two particular groups of representative payees: 36 people appointed by courts in some guardianship position and 17 people who ran group homes. Table 5-1 shows the results of the reinterview study by type of payee.

The first group of court-appointed conservators and legal guardians were likely to have organized documentation of benefit expenditures. Many of them employed administrative staff who did the actual record-keeping. The reinterviewers had expected that misuse might be minimal among this group given the strict requirements of state courts. Many states have strict rules about how guardians and conservators must keep records, and require mandatory training. Some of the attorneys in this group noted that a benefit of accepting court-appointed payee positions was to maintain a positive relationship with judges. This information led the committee to consider that accountability is probably driven by the courts rather than by SSA.

However, misuse did occur even within this group. A case categorized by the committee as misuse involved a lawyer appointed by a probate court.

TABLE 5-1 Misuse by Type of Payee in Reinterview Cases

Type of Representative Payee	No Misuse	Misuse	Uncertain or Potential Misuse	Total
Court appointed	27	2	7	36
Group home	6	7	4	17
All other	10	7	6	23
Total	43	16	17	76

SOURCE: Data from the reinterview of selected respondents to the national survey of representative payees and beneficiaries conducted for the National Academies Committee on Social Security Representative Payees (2006).

At the time of the reinterview, he was the payee for ten beneficiaries. He did not use direct deposit. He did not seem to know much about the beneficiaries he served; indeed, SSA had to notify him that one of them was in jail. What triggered the committee's interest was that there was one $600 withdrawal from a beneficiary's account apparently paid to the representative payee himself. He said yes, that he had taken it because otherwise the state (or maybe SSA) would take it, since the amount in the account exceeded what SSA allowed. He did not use the money for the beneficiary and was quite clear that he used it for himself to avoid having the state take it.

Another case in this group was that of a lawyer who was the payee for his adult child. The beneficiary worked part time at a relative's law firm, for which he received a W-2 form. This salary went into his account, along with his SSA funds. The beneficiary's expenditures included rent, utilities, club membership, computer lease, allowance, medical expenses, and renter's insurance. In the past, the payee had had to pay back a large amount to the SSA for overpayments. Several entries in the check register were to the payee, who claimed to have receipts for all of them, but did not show them to the interview team. The committee categorized this as a case of uncertain misuse.

For the second group, 17 people who ran group homes, SSA is often unaware that payees run such homes. The differences among states concerning regulation of these group homes accounted for much of the misuse found among these payees: For example, Oregon and Florida appear to have fairly rigorous standards for group homes (see Chapter 6). There were varied patterns of misuse among this group.

In one case, a woman took care of three elderly beneficiaries in her home. She commingled all funds and used the money for food, clothing, cleaning supplies, medications, taking the beneficiaries out to dinner once a week, and her own car maintenance. She did not keep separate accounts, nor did she keep records of expenditures. The funds were used to keep the group, including her, afloat. The committee characterized this case as misuse.

In another situation, a group of related payees in one state ran several homes for the mentally handicapped. They refused to use direct deposit and pooled all beneficiary funds. There was no rationale for the amounts charged for room and board, which were very high. Payees in this family learned from each other how to set up these homes and they applied for payee status at different SSA offices, even though they lived close to each other. These payees met together on a regular basis to discuss fees and other policies for their group homes. These cases were characterized as misuse by the committee.

These examples show the potential for problems when creditors of beneficiaries also serve as representative payees. Yet at least one state requires a group home administrator or owner to be the representative payee for all of the beneficiaries under that person's care, although this state also requires separate accounts and very specific record-keeping, as well as training and standard job descriptions (see Chapter 6).

Among the 23 "other" cases, there was also a great deal of misuse and potential misuse. In one case, the payee admitted that he charged each of his beneficiaries $25 a month to be their payee. In another case, the payee commingled the beneficiary's funds with her own bank account, kept no records, and kept track of the balance "in her head." There was no documentation or evidence of how the money was spent.

In addition to a relatively high rate of misuse and potential misuse for the 76 cases the committee examined in depth, the committee is concerned about the implications of the high rate of no-shows and nonrespondent cases. As noted above, the interviewers could not locate payees in four of the cases, by registered mail or by phone. One can only wonder how they receive beneficiary checks and whether they interact with their beneficiaries. In the other 19 cases, in which the payees had received the letter and talked to someone on the phone, the payees still did not show up at their appointed times, despite such accommodations as rearranging appointments. There could be several explanations for such behavior. Although the letter stated that the reinterview was mandatory, the payees may have believed from past experience that SSA would not penalize them for failing to attend. Moreover, this group contained a high percentage of felons and people who had served time in prison, so they may have been wary of responding to government inquiries. Another possible reason is that since a higher percentage of these people than all payees were under 50, they were more likely to be employed, and it may have been difficult for them to take time off from their jobs. Indeed, some payees who did participate in the reinterviews complained that it was costing them money. The committee was not able to further investigate the reasons for the high number of no-shows, and we are concerned that failure to participate could indicate a substantial number of additional misusers.

The committee's analysis of the 76 payees who were reinterviewed cannot be used as a basis for determining what percentage of representative payees as a whole are misusers, because we did not include representative payees with low scores on the 15 variables in the reinterview sample. If the IG selected a sample for audit that included cases from all sampled domain groups identified in Chapter 2, an overall estimate of misuse could be made. However, the committee can say with some assurance that misuse among payees with high scores on the 15 variables is much more common than that detected by other methods currently used by SSA to detect misuse.

Indeed, misuse is probably more common than we found through the reinterview process. For example, mothers of beneficiary children were not included in the reinterview process, but this group was described in the in-depth study as the group with the largest number of misusers. In Chapter 4 we note that there were nine representative payees who did not know where their beneficiaries were; it is likely that at least some of these cases involve misuse.

The large number of cases in the uncertain or potential misuse category also illustrates how difficult it is to identify misuse with any certainty under the current definition. Many of these uncertain cases were so classified because the payees did not provide any financial records or did not have sufficient records, thus preventing any conclusive determinations as to misuse. Because SSA does not assess penalties for failure to keep records, some payees know that they can fail to identify expenditures. Finally, even in the cases that we classified as "no misuse," there were many violations of SSA policies. This, too, is important because violations may cover up misuse.

Some of the more common violations are (1) commingling of funds, (2) failure to notify SSA of changes in a beneficiary's situation, (3) failure to maintain clear and separate accounts, and (4) failure to make financial records available when requested by SSA. In the next section we discuss these violations, using examples from the reinterviews for illustration.

Conclusions and Recommendations

Through the committee's analysis, we learned that many payees were sincerely interested in the welfare of their beneficiaries, but we also learned that there were unscrupulous people taking advantage of some beneficiaries. And we learned that some payees just do not seem to understand the policies and inadvertently misuse funds. Even in our small sample of reinterviews, we found some very sad cases in which some payees use beneficiary funds to keep a household afloat.

The committee concludes that its strategy in attempting to identify characteristics associated with misuse was successful. Three of the characteristics—being a felon or having served prison time (characteristics 8 and 9) and receiving welfare benefits (TANF or its predecessors, characteristic 5)—were most often blank in the RPS, so their usefulness is limited until SSA undertakes an effort to ensure that these data and updates are obtained. The standard errors of the estimates are high and, therefore, no one variable can be declared a good discriminator.

Using the characteristics led to the identification of a substantial number of court-appointed representative payees. Administrative data in the RPS could be used to identify these payees and eliminate them from the process. It also seems inefficient to review cases for which the benefits are a very small amount, perhaps less than $500.

CONCLUSION The characteristics identified in the committee's in-depth study of misuse are effective in targeting representative payees for auditing for the purpose of detecting misuse.

RECOMMENDATION 5.1 The Social Security Administration (including its Inspector General) should use probability sampling with targeted sample selection, using criteria associated with misuse of funds such as those suggested in this report, to audit representative payees who are more likely to be misusers.

RECOMMENDATION 5.2 The Social Security Administration should develop criteria associated with misuse of funds, such as those suggested in this report, to select and monitor representative payees.

In the process of selecting new representative payees, SSA could add the scores and try to avoid selecting people with high scores. In the monitoring process, the SSA could set up a differential program of monitoring, giving more frequent attention to payees with higher scores than average.

CONCLUSION The use of a specialized team of auditors was effective in uncovering misuse of funds by representative payees.

RECOMMENDATION 5.3 The Social Security Administration should establish a team of experts, such as the audit teams used in the committee's study, to audit those payees who are suspected of misuse or who have been included in a targeted sample of potential misusers.

Mothers of minor children are 54.5 percent of all representative payees and were the most frequent misusers found in the in-depth study of misuse (see Appendix D). The above list of characteristics would not identify mothers. Therefore, the committee suggests that another set of characteristics focused on parents, perhaps those with frequent changes of custody, could identify a sample that should be monitored more closely.

POLICY VIOLATIONS

Commingling of Funds

One of the most common violations of SSA policy we found was the commingling of funds by the payee. This could happen in at least two ways. One is when the representative payee had a bank account for the beneficiary into which all money for the beneficiary was deposited. That might mean that money from investments, alimony, Department of Veterans Affairs benefits, and other sources were mixed together, as frequently happened when the payee was an attorney, conservator, or a professional guardian.

In the booklet that SSA gives to representative payees (see Chapter 1), commingling is prohibited (U.S. Social Security Administration, 2006). Notwithstanding that stated prohibition, SSA, when asked about this by the committee, replied that it has no restriction on holding SSA funds with other funds in the same account as long as a clear record of expenditures is kept: "SSA has no restriction regarding holding Social Security funds along with other funds in the same account" (see Appendix C).

When commingling is done by guardians who are also representative payees, it allows a guardian to calculate his or her fees based, in part, on SSA funds (because such fees are allowed by state courts to be based on a percentage of the total income for the beneficiary). However, during the reinterviews, the payees usually said the fee was paid only from the non-SSA funds. There was also evidence during the reinterviews that other representative payees commingled funds, but without the type of court oversight given to those payees who were also appointed guardians or conservators.

A second way funds were commingled was when a representative payee deposited funds for several people in a common bank account. In some cases, a payee with only one beneficiary put the SSA funds into his or her own bank account. In other cases, a payee with several beneficiaries put the SSA funds for all of them into a common account. In these latter cases, all expenses for the upkeep of the group came from these pooled funds, and there was little documentation of expenditures available. The money from any one beneficiary was going to support others, sometimes a family. The committee deemed these cases to be misuse. In some of these cases, when a beneficiary went to jail, the payee failed to notify SSA, probably because the household could not survive without that money. Pooling of SSA benefits by a representative payee, either with his or her own funds or with the funds of others, made it extremely difficult to determine whether the SSA benefits were used solely for the specified beneficiary.

Failure to Notify SSA of Changes in a Beneficiary's Situation

SSA rules about the role of representative payees state that a payee must notify SSA when there are changes in the beneficiary's status that might affect benefit levels. The changes might include death, a job, an inheritance, or an incarceration. When a beneficiary dies, it is incumbent on the payee to notify SSA and return any check received for that month. Similarly, when a beneficiary is institutionalized, the payee is required to notify SSA. While a beneficiary is in jail or some other kind of penal institution, he or she is not entitled to receive SSA benefits. Likewise, when a beneficiary obtains a job and is earning money, the payee is to notify SSA, since the income from the job may affect the beneficiary's rights to collect benefits or the amount of the benefits. In several cases in the reinterview sample, payees had not notified SSA of these kinds of changes.

Failure to Maintain Clear and Separate Accounts

One of the duties of a payee is to maintain clear and separate accounts that provide a clear audit trail of all financial transactions for the beneficiary. This duty was not met by a large number of payees in the reinterview sample. One commonly heard justification was that the amount received was too small and so it would have been unduly costly and burdensome to open a separate account. Others suggested that because the SSA amount was a small percentage of the beneficiary's total income, it would be costly to separate it out. These same representative payees suggested that since most of the monthly income was going to pay for some kind of institutional care (nursing home or assisted living) and the costs of this care were greater than the benefit amount, a separate account was not needed. Some payees said that they did not know that they should be keeping separate accounts. Others seemed to be baffled by the record-keeping requirements. Yet we note that many payees kept excellent records, especially the lawyers, conservators, and professional guardians in states in which the courts examined the records. Some of the guardian payees did not know exactly what the SSA program required in terms of records, but knew it was less than what the courts asked for. Those whose record-keeping was poor often fell into the pool of cases for which the committee thought that misuse was uncertain. Without a record of expenditures, it was difficult to determine whether misuse had occurred.

Failure to Make Financial Records Available when Requested by SSA

Many payees did not bring any records to the reinterviews, even though they had been specifically told to do so by certified mail and by telephone. The IG's report also mentioned that records were not always available.

It is apparent that SSA does little to enforce its policies and that the payees are aware of this lack. In fact, one payee called the SSA a "toothless tiger" because it did not contact payees often, if at all, and because the only accounting to be made was the very basic annual accounting form.

Some of the violations noted in this section led to the committee's judgment of misuse; others did not. The committee assigned the "misuse" label only when there was clear evidence that the beneficiary's funds were spent on others. The committee assigned the label of "uncertain/potential misuse" when there were few or no accounting records, SSA benefit amounts significantly exceeded the beneficiary's apparent immediate needs, or some very large expenses were claimed that could not be verified.

ANNUAL ACCOUNTING FORM

In addition to offering the targeting tool to detect misuse, the committee proposes revisions to the annual accounting form. As discussed above, the purpose of the annual accounting form is to allow SSA to monitor the payee performance and to detect misuse. Although accounting forms are to be completed annually by all payees, only about 60 percent of the accounting forms are returned by payees after the first contact and scanned into the system with no further action needed.

The committee noted several other flaws with the accounting process. First, the process is not set up to yield data for research or analysis. For example, SSA does not calculate mail return rates, keep track of item nonresponse, nor use the data in any way for statistical analysis. Because only images of the forms are stored, no individual data items can be retrieved from the system. Second, there is no quality assurance program associated with the accounting system: no one checks the work of the clerks who enter the data. Third, the annual accounting forms that are sent back to field offices for follow-up are often not returned for storage in the SSA processing center (in Wilkes-Barre) when the cases are resolved. Consequently, any accounting forms that were initially problematic are not retrievable for future analysis. The IG's report noted that about 28 percent of the forms requested were not available.

Throughout our field site visits, SSA staff told us that the annual accounting form, in its current version, is not an effective tool for detecting misuse. It is too easy for representative payees to learn that if they just fill out the form with some plausible but not necessarily accurate information, there will be no follow-up or other consequences. Essentially, the current monitoring process is an "empty threat" that can easily be subverted.

The committee concurs with the information it received from SSA staff and representative payees that the current annual accounting form is a useless, expensive administrative tool that does not yield the sort of data that is necessary to uncover misuse. Beyond meeting a 20-year-old court mandate to monitor all representative payees, in its present format the accounting form provides a very limited return for the time and money invested in handling the paperwork.

CONCLUSION The current annual accounting form is of limited value in monitoring representative payees for potential misuse.

RECOMMENDATION 5.4 The Social Security Administration should redesign the annual accounting form to obtain meaningful accounting data and payee characteristics that would facilitate evaluation of risk factors and payee performance.

A preliminary attempt at redesign of the annual accounting form by the committee (but without cognitive testing) is presented in Appendix F. Our goal was to provide an example of what could be done to improve the form. We added questions that incorporate the characteristics of misuse that we have identified. We revised the form to omit providing to payee the total amount of benefits SSA has sent, thus reinforcing the requirement that payees keep (and consult) records. We added questions that reinforce the duties of the representative payee to avoid the policy violations that we discovered in our work. Most importantly, we propose that online entry be offered on the Web and that paper forms are scanned using optical character recognition to eliminate most clerical review of forms and to facilitate statistical analysis and data checking. We stress, however, that these suggested changes—or any others—need to be cognitively tested to make sure the language on the form is as clear as possible and suitable for payees of all educational levels.

No form, by itself, is going to detect program misuse. However, if a form can be used to obtain information on characteristics of interest, it could then be combined with a rigorous program of audits such as the one the committee developed.

BANK-ACCOUNT-LINKED DEBIT CARDS

In 2002 the Social Security Advisory Board asked SSA's IG to develop an issue paper on the use of stored value cards (SVC) by representative payees. The IG's report (U.S. Social Security Administration, 2002) argued that there would be six potential benefits to the use of such cards for both the ease of distributing benefits and the straightforwardness of monitoring expenditures:

1. Tracking representative payees' purchases—an SVC would automatically keep an electronic record of most purchases.
2. Timely identification of representative payee spending—SSA could obtain spending information from financial institutions on the type and amount of many payees' expenses. This information could be available electronically and be reviewed at various intervals (weekly, monthly, and annually).
3. Identification of questionable expenses—records from an SVC can be used to create exception reports, which can allow SSA to iden-

tify quickly instances of questionable expenses, unusual spending patterns, and program ineligibility. Merchant blocking could be used to prevent payees from making purchases with certain vendors, which is done in the food stamp program (and also includes prohibited items).

4. Identification of conserved funds—tracking balances on SVCs would permit SSA to identify quickly Supplemental Security Income program ineligibility when balances exceed $2,000.
5. Transferring funds when payees change—with an SVC, a beneficiary's balance could be electronically transferred to a new payee; SSA would not have to wait for the former payee to return any balance to the SSA.
6. Elimination of the paper accounting form—use of SVCs might provide administrative savings of the cost of mailing and processing the annual accounting forms.

SSA did not accept the IG's suggestion to carry out a test of the use of SVCs. However, it is now 5 years later, and there is experience with many other federal agencies using such cards successfully. Bank-account-linked debit cards can clearly provide protections and ease for representative payees for Social Security beneficiaries.

In addition, legislation adopted in 1996 provides that low-income recipients of federal payments should be able to easily open and maintain bank accounts through direct deposits, and limits charges that the bank could impose.[2] Although there are numerous types of stored value cards available in the market place (e.g., gift cards), most are not attached to bank accounts and as such, they do not offer the same types of consumer protections that bank accounts do under federal law. These types of bank-account-linked debit cards are also linked to congressionally established low-income bank accounts known as electronic transfer accounts (ETAs).[3]

An ETA account, as designed by the Treasury Department, ensures that individuals who receive federal payments electronically have access to an account at a reasonable cost and with the same consumer protections available to other account holders at the same financial institution. These protections, most important to beneficiaries and their representative

---

[2] The Debt Collection Improvement Act of 1996 (DCIA), Public Law 104-134, Omnibus Consolidated Rescission and Appropriations Act of 1996, Chapter X, Sec. 31001 (amending 31 U.S.C. § 3332), outlines the duties of financial institutions designated as Financial Agents and ensures compliance with the ETA standards.

[3] An ETA is a low-cost account that is made available by participating federally insured financial institutions to individuals who receive federal benefit, wage, salary, or retirement payments. The account allows recipients to receive federal payments electronically in accordance with the electronic funds transfer provision of the DCIA.

payees, include: disclosure requirements, a dispute resolution mechanism, a requirement that charges be documented, protection from unauthorized withdrawals, an absolute barrier to attachment of the deposits coming from the federal government, and guarantees of limited fees on the account.

Of particular interest for representative payees, the accounts must have the following characteristics:

- Be an individually owned account at a federally insured financial institution.
- Accept electronic federal benefit, wage, salary, and retirement payments and such other deposits as a financial institution agrees to permit.
- Be subject to a maximum price of $3.00 per month.
- Have a minimum of four cash withdrawals and four balance inquiries per month, to be included in the monthly fee, through any combination of proprietary automated teller machine transactions or over-the-counter transactions.
- Provide the same consumer protections that are available to other account holders at the financial institution.
- Allow access to the financial institution's online point-of-sale network, if any.
- Require no minimum balance, except as required by federal or state law.
- Be either an interest-bearing or noninterest-bearing account, at the option of the financial institution.
- Provide a monthly statement.

The beneficial elements outlined by the IG's 2002 report would be present in ETA accounts that include a bank card with the additional consumer protections established with the Bank Act of 1996. Using these types of accounts would provide advantages for both the ease of distributing benefits and the straightforwardness of monitoring expenditures.

If the SSA encouraged the development and expansion of ETAs for the use of payees who serve a single beneficiary, it is not unreasonable to foresee the day SSA could waive the costly and mostly useless annual reporting requirement and simply collect the monthly statements made available through banks. Nearly instant monitoring would be available, along with a database for statistical analyses to discover potential misuse.

RECOMMENDATION 5.5 The Social Security Administration should conduct a test of bank-account-linked debit cards for representative payees.

RESEARCH AND DEVELOPMENT STAFF

The Representative Payee Program does not have a staff dedicated to research on new methodologies for detecting misuse or for the development of computer and database systems that would enhance the program and increase field staff efficiency. In the field offices, there are no staff whose priority job is to recruit, train, and monitor representative payees; to investigate potential cases of misuse; to train claims representatives in how to use the RPS; to systematically monitor accounting requirements and lump-sum payments; to monitor changes in custody of minor children; to serve as a resource for questions from payees and beneficiaries about payee responsibilities; or how to use criteria associated with misuse of funds by payees. If the identification of misuse is made easier—with a revised annual accounting form that helps identify misuse, or by bank-account-linked debit cards, program quality and cost-effectiveness would be enhanced.

RECOMMENDATION 5.6 The Social Security Administration should initiate a research, development, and support function for the Representative Payee Program to promote quality and cost-effectiveness in its operations.

6

Program Policies and Practices

his chapter addresses the second specific item in the committee's charge: "learn whether the representative payment policies are practical and appropriate." After a brief discussion of a systems perspective of the program, there are five major sections that correspond to stages in the representative payee appointment process: establishing the need for a representative payee, selection of payees, training and support for payees, monitoring and accountability, and termination and transitions. The final major section of the chapter considers state laws and programs that are relevant to the Representative Payee Program, and we end with observations about the variation in the management of local offices.

A SYSTEMS PERSPECTIVE

The Representative Payee Program involves a complex interactive system of people, policies, and procedures. As such, the performance of the program is best viewed as the outcome of many interactions among Social Security Administration (SSA) staff, payees, beneficiaries, and, on occasion, other people, such as beneficiaries' family members and legal guardians and conservators (who are not payees). We therefore adopted a systems perspective in our assessment of program policies and practices by examining each step in the process in terms of the (1) interactions that occur, (2) the factors that influence those interactions, and (3) the effect on service to beneficiaries.

The Representative Payee Program is dynamic, and it is influenced by a complicated environment of policies and interactions between the beneficiaries and their representative payees. The amount of the Social Security benefit and the circumstances of the beneficiaries (e.g., health status, living situation, availability of additional beneficiary income or resources) may also affect the relationship between beneficiaries and representative payees. Figure 6-1 depicts a systems view of the program, in which many factors, interactions, and policies influence the implementation and the outcome of each stage. In Figure 6-1 these factors are shown to the left of the process schematic. To understand the complexity of the system, consider the first stage of the process, establishing a beneficiary's need for a representative payee. Program policies, promulgated publicly through the SSA's Program Operations Manual System (POMS), establishes protocol and guides SSA claims representatives through the determination process. In addition to the formal policies, however, the effectiveness of claims representatives is also influenced by their training and experience and by the impact of SSA management policy on their motivation and performance (e.g., rewards for productivity, effectiveness). Beneficiary characteristics are critical as well (e.g., lucidity, cognitive ability, institutionalization, mental disability status, health), and input from a beneficiary's family can come into play. The determination made by the Disability Determination Service office can influence a claims representative's determination of need. Even a state court's assignment of a legal guardian or conservator for the beneficiary must be a consideration for a claims representative. Overall, a wide variety of factors both internal and external to SSA affect the determination of need of a representative, and similarly complex scenarios occur with all the other stages of the program.

The rest of this chapter evaluates SSA policies and practices for each stage of the process using data from the committee's survey, the site visits, and SSA administrative data. The conclusions and recommendations are included in the section for each stage of the process.

ESTABLISHING THE NEED FOR A REPRESENTATIVE PAYEE

Process for Determining Need

The SSA has broad responsibility and discretionary authority to determine whether to pay benefits directly to the beneficiary or to pay a representative payee on the beneficiary's behalf (42 U.S.C. §§ 405(j), 1007, & 1383(a)(2)). The law and regulations specify that a payee should be appointed for

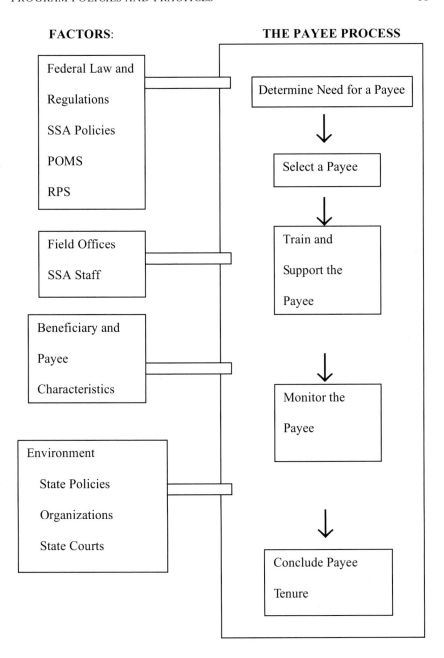

FIGURE 6-1 Systemic View of the SSA Representative Payee Process.

- beneficiaries under the age of 18,
- beneficiaries whom state courts find are lacking capacity, and
- beneficiaries determined by SSA to need a payee.

In assessing a beneficiary's capability to manage or direct the management of benefit payments, federal regulations stipulate that court determinations, medical evidence, and any other helpful information may be considered. SSA's operating policies in the POMS emphasize the principle that adult beneficiaries are deemed capable unless there is evidence to the contrary. These policies require SSA to consider the beneficiary's ability to meet daily needs and manage money.

In this section we discuss the effectiveness of this flexible system of protocols to evaluate the need for a representative payee, particularly the extent to which payees are appointed for beneficiaries who are able to manage their own payments and the types of conditions that lead SSA to determine there is a need for a payee.

Accuracy of Need Determinations

In the committee's field office site visits, some staff said that some beneficiaries with payees do not need them. This situation can result from either a prior temporary need for assistance that is no longer applicable, for example, due to a beneficiary's short-term convalescence or mental incapacity, or because the initial determination of need may simply have been incorrect. The POMS provides a set of questions for a claims representative to use as a guide in determining a beneficiary's capability of managing his or her financial benefits. The POMS further states that if there is no legal determination or medical evidence to establish the lack of capability, then SSA staff must document the need for a payee through the use of lay evidence (POMS GN 00502.030).

Estimates based on our survey of beneficiaries who are 18 years or older suggest that the problem of incorrectly appointing a payee is relatively minor. The survey showed that beneficiaries and payees agreed the beneficiaries could manage their payments on their own in only 4.4 percent (0.7)[1] of cases (see Table 6-1). The percentage of beneficiaries who agreed with their payees that they could manage their own funds was higher for beneficiaries who were 18-64 years old, who were parents of payees, who

---

[1]For all estimates in this chapter, the number in parentheses following the estimate is the standard error of the estimate. As a general rule, users can approximate a 95-percent confidence interval for the estimate by adding and subtracting two standard errors to the estimate. When two estimates have confidence intervals that overlap, the two estimates are not statistically different at the .05 level of significance.

TABLE 6-1 Representative and Beneficiary Agreement on Ability of Beneficiary to Manage Own Benefits, for Beneficiaries 18 Years or Older

Beneficiary or Payee	N	Percent (SE)[a]	Number of Beneficiaries (SE)
Beneficiary			
All beneficiaries	1,402	4.4 (0.8)	78,800 (13,900)
18 to 64 years old	1,037	4.9 (0.9)	71,600 (13,100)
65 years or older	365	2.2 (1.3)	7,100 (4,400)
Payee			
Parent of beneficiary	295	3.0 (1.1)	25,100 (9,600)
Child of beneficiary	162	9.0 (3.4)	19,900 (8,300)
Other relative	268	4.6 (1.8)	23,700 (9,100)
Nonrelative	548	7.6 (1.1)	10,100 (1,400)
Lives with beneficiary	651	2.0 (0.8)	23,000 (9,200)
Lives separately from beneficiary	751	8.6 (2.0)	55,700 (13,100)
Individual with one to four beneficiaries	1,170	4.7 (0.8)	78,600 (13,900)
Individual with five or more beneficiaries	98	0.3 (0.3)	100 (100)
Organization	147	0	0

[a]For all estimates in this chapter, the number in parentheses following the estimate is the standard error of the estimate. As a general rule, users can approximate a 95-percent confidence interval for the estimate by adding and subtracting two standard errors to the estimate. When two estimates have confidence intervals that overlap, the two estimates are not statistically different at the .05 level of significance.

SOURCE: Data from the national survey of representative payees and beneficiaries conducted for the National Academies Committee on Social Security Representative Payees (2006).

were unrelated to their payees, and who did not live with their payees (see Table 6-1). At the same time, 8.3 percent (0.7) of payees would like to talk to SSA about their beneficiaries' managing their own funds.

Although the prevalence of what may be erroneous payee assignment is relatively low, there was broad interest among field office staff in establishing a more effective relationship between the Disability Determination Service offices and the SSA field offices. At issue were perceived disparities between the assessments by SSA claims representatives of the need for payees and the evaluations of the Disability Determination Service offices of disability and how that evaluation related to the need for a payee. The committee agrees with many field staff that a joint determination process may lead to better assessments. We also agree with what we heard from many in the field that there is also a need to check with other community agencies and reporting authorities about the credibility of potential payees.

Occasionally, there are beneficiaries who are assigned a temporary payee, or who are temporarily put into direct payment status despite needing a payee, e.g., because no payee can be identified or the payee has died or been terminated. These situations should be closely monitored or consideration should be given to setting up a process to regularly evaluate the continuation of the payee status and beneficiary capability.

Another potential error in the process involves undetected need: the situation in which a beneficiary without a payee needs one. However, the scope of our study did not include beneficiaries without payees assigned to them, and thus we cannot estimate the number of beneficiaries in need of a payee but currently lacking one.

CONCLUSION A small fraction of beneficiaries aged 18 or older are potentially capable of managing their own payments. This is an acceptable level of accuracy in determining beneficiary need for a representative payee, given the complexity of the determination task and the size of the system.

SELECTION

Recruitment

Selecting appropriate payees is a critical factor in meeting beneficiaries' needs to manage their SSA payments. The committee's site visits to SSA field offices indicated that current definitions and policies in the POMS adequately describe desirable characteristics of a payee. However, inconsistencies were reported in the criteria used for payee selection. For instance, one staff member reported augmenting minimal requirements in POMS with criteria that focus on the personal relationship between a payee candidate and the beneficiary. Other "additional" characteristics reported by staff as selection criteria are shown in Box 6-1. Some staff also attempt to tailor payee selection to the specific needs and situation of a beneficiary, such as the beneficiary's living arrangements or evidence of beneficiary alcohol or substance abuse.

Many of these factors appear in the POMS preference lists for minor beneficiaries, adult beneficiaries without substance abuse problems, and adult beneficiaries with substance abuse problems (see below). The lists are tailored to specific beneficiary types and establish (but do not require) a recommended priority for selecting a payee. With regard to recruitment of payees, it appears that different claims representatives emphasize different factors in the POMS preferences lists and use them as requirements rather than preferences.

It is also worth noting that some claims representatives used an intangible criterion—intuition, a "gut feeling"—in the selection of a payee.

**BOX 6-1
Criteria for Identifying Payee Candidates
as Reported by SSA Field Staff**

Is financially capable
Understands that the primary job is to see that the money is spent for the beneficiary's needs
Can manage his or her money
Has enough financial savvy to know how to keep beneficiary funds separate from own funds
Knows how to manage money

Has a close relationship with beneficiary
Has frequent contact with the beneficiary
Is most likely a family member (e.g., parent, spouse, other)
Shows concern for the beneficiary
Possesses an in-depth knowledge of the beneficiary
Knows the beneficiary—at least first and last name, perhaps also date of birth
Lives with the beneficiary
Lives very close to the beneficiary
Can check in on the beneficiary once a day
Knows what is going on with the beneficiary

Has no or very few risk factors for potential to misuse funds
Should not be a felon—but if it happened twenty years ago, it is probably okay now
Is gainfully employed or has his or her own sources of income
Has no criminal record
Is not a fugitive felon
Is not owed money by the beneficiary
Has a stable employment record
Has a high school or equivalency diploma
Has had no encounter with law enforcement
Has no record of alcohol or drug abuse

The meeting between an SSA staff person and a payee candidate affords an important opportunity for the staff person to interact with the payee candidate and observe physical and nonverbal cues (e.g., nervousness, cognitive ability, coherence, physical condition, potential drug or alcohol use, sincerity). Especially among the seasoned veteran staff, the personal interview provided them with a sense of whether or not a specific candidate would be appropriate as a particular beneficiary's payee.

The survey defined payees who were related to their beneficiaries as blood relatives, relatives by marriage or marriage-like partnerships, and close friends. The vast majority of beneficiaries were related to their payees, 94.7 percent (0.1). The selection process is likely to be relatively easy when beneficiaries know related persons who are willing to serve as payees. However, for adult beneficiaries who did not have related persons available to act as their payees—4.7 percent (0.1) or 214,300 (4,400) beneficiaries—the payee identification and selection process can be more challenging.

Among adult beneficiaries with unrelated payees, almost one-half did not personally ask someone to be their payees, 43.9 percent (2.5); their payees were obtained through other means. Families were not involved in the selection decision for nearly one-half of adult beneficiaries with nonrelative payees, 43.5 percent (2.6). SSA was responsible for identifying and recruiting the payees for about one in six adult beneficiaries with unrelated payees, 14.8 percent (2.6). For nearly two-thirds of beneficiaries with unrelated payees, the payees had volunteered to serve them. Thus, SSA had to recruit and secure payees for more than one-third of beneficiaries with nonrelative payees, 35.2 percent (2.4). These results give rise to concern about the amount of time needed by SSA staff to identify appropriate payees as well as the need for tighter links to organizations that offer volunteer payees to serve beneficiaries with challenging circumstances (such as the absence of a relative to act as payee).

To handle the more difficult cases of finding appropriate payees, SSA field offices are required by the Social Security Act (Section 205(j)(3)(G)) to keep a list of local payee sources. In addition, the POMS (GN 00502.100) encourage field offices to develop ongoing, cooperative relationships with community social service providers who can often provide payee contacts. However, the committee's site visits suggest that staff resources for carrying out these required and recommended activities are severely limited: we found no such lists of payee sources in the field offices, and there appeared to be few strong community relationships with social service providers.

Another source of support for the selection process comes from nongovernmental agencies and organizations that train and monitor volunteer payees. Such services have been or are currently being provided by national nonprofit organizations as well as local community organizations. For example, the AARP Foundation began a volunteer representative payee trainee program in 1981; in 2004, approximately 5,000 beneficiaries were served nationally by this program (AARP, 2005). However, an organization must be willing to contribute funds as well as staff to coordinate a volunteer payee program, and many local community and national organizations cannot support such activities in perpetuity. SSA investment in promoting such programs may improve the supply of volunteer or community payees, particularly for handling difficult cases.

SSA staff appear to be tailoring the payee selection to the specific needs and situation of a beneficiary, which is ultimately advantageous to beneficiaries. Although flexibility is appropriately part of the payee selection process, there is significant variation in how local staff apply criteria to select payees, both within and between field offices. For example, the recommended *preferences* provided by POMS are often used as *requirements*. Although such efforts to improve the selection of payees can be considered laudatory, they also suggest a need to standardize the payee selection process.

SSA policy requires the maintenance of a source of payee volunteers and community organizations with payee volunteer programs to address situations in which payee candidates are not readily available for a given beneficiary in need. However, in some local offices such volunteer pools were limited or did not exist, and staff resources for developing them were insufficient. The lack of resources for ensuring the development and maintenance of payee volunteer pools severely hampers the program in some offices.

RECOMMENDATION 6.1 To help mitigate shortages of payees, the Social Security Administration should create a program to identify, train, certify, and maintain a pool of voluntary, temporary payees that are available on an as-needed basis. If such a program is authorized, the Social Security Administration should work with and obtain help from the courts and volunteer organizations in designing it.

Recruitment of Payees for At-Risk Beneficiaries

At-risk beneficiaries are those with precarious health, behavioral, or living situations. Such beneficiaries include but are not limited to those who are homeless, have alcohol or substance abuse problems, are mentally unstable, or have chronic health conditions that require special living or treatment arrangements. SSA staff expressed concern with recruiting payees for such at-risk beneficiaries, noting that the needs of these beneficiaries far exceed the fiduciary service that could be provided by an individual payee or by SSA. Staff told the committee that at-risk beneficiaries deserve special payee arrangements that currently do not exist in the system. For instance, homeless beneficiaries have daily needs of food and shelter, yet their benefits are either insufficient, unavailable, or are not being used for those needs. Mentally ill beneficiaries are often left to fend for themselves in the community and would benefit from an organizational assisted living arrangement, yet affordable choices are not always available to them. A third example concerns substance abusers. These beneficiaries often seek to maximize their cash allocations from payees—to the detriment of basic needs—in support of their addiction.

The POMS provides detailed guidelines for the selection of a payee for certain classes of at-risk beneficiaries, such as known alcohol- and substance-abusing beneficiaries (POMS GN 00502.105).

> Select the best payee available from this list of preferred applicants: a community-based nonprofit social service agency bonded and licensed (if required) by the State; a Federal, State or local government agency whose mission is to carry out income maintenance, social service, or health care-related activities; a State or local government agency with fiduciary responsibilities; a designee of an agency (other than of a Federal agency) referred to above, if appropriate; or a family member. When none of the preferred payees above are available, select the best payee available from this list of alternate sources: a legal guardian with custody who shows strong concern for the beneficiary's well-being; a relative or friend with custody who shows strong concern for the beneficiary's well-being; a public or nonprofit agency or institution with custody; a private, for-profit institution with custody and is licensed under State law; or anyone not listed above who is qualified and able to act as payee, and who is willing to do so; an organization that charges a fee for its service.

A representative payee's responsibilities for these and other at-risk beneficiaries require special training and professionalism that transcends the current model of a "suitable representative payee," as well as the guidance provided by the POMS. SSA staff reported that finding payees for at-risk beneficiaries tends to be the most challenging selection task. Such payees tend to be nonrelative friends or acquaintances, and payee tenure tends to be short in comparison with the tenure of payees for other beneficiary types. Finding appropriate payees for at-risk beneficiaries is perceived as a significant challenge, and at-risk beneficiaries are believed to be at higher risk for payees' misuse of funds.

The use of payees from volunteer groups or for emergency situations was discussed in some of the committee's site visits. As mentioned above, SSA staff are sometimes unable to identify suitable payees despite their charge to maintain a pool of such candidates. Staff reported to us that sometimes a less-than-optimal candidate is selected simply because there is no other choice, and the beneficiary's needs for a payee are urgent. For such situations, the staff suggested the development of a pool of trained, willing payees that could be tapped as needed for emergency or other temporary situations.

Some field office staff said that they believe that at-risk beneficiaries are better served by fee-for-service payees who are professionals and may be better acquainted with the payee system and rules. Yet these types of payees are at the end of the POMS list of "preferred" payees for substance-abusing beneficiaries who need payees. Staff noted that although small group institutions that provide high-quality care are often better suited than individuals to serve at-risk beneficiaries who suffer from substance or alcohol abuse or have mental disabilities, the bonding requirement can be too high a burden for these small commercial payees.

Finally, the changing demographics of the United States are worth mentioning here. The number of beneficiaries with mental illness, mental retardation, and developmental disabilities; the homeless; and persons in the end stages of HIV/AIDS will continue to increase in the coming years (see Teaster, 2003). Thus, the problem of identifying payees for at-risk beneficiaries can be expected to increase and worsen if the current system is not changed.

CONCLUSION It is difficult to find appropriate payees for at-risk beneficiaries. Fee-for-service payees may be better for at-risk beneficiaries because they are professionals and may be licensed and are better equipped to deal with situations posed by at-risk beneficiaries.

RECOMMENDATION 6.2 Congress should authorize the Social Security Administration to expand the fee-for-service part of the program to include appropriate small organizations and individuals who are willing to serve as payees for at-risk beneficiaries: people with mental illness, alcohol or substance abuse problems, severe disabilities, and those who are homeless.

Suitability

Once a representative payee candidate is identified, SSA must determine the person's suitability to serve as a payee. The process of determining suitability involves

- completion of an application by the candidate;
- verification of candidate's identity;
- assessment of the candidate's exclusion factors (see below); and
- in most but not all cases, a personal interview with an SSA field office staff person.

SSA policy (in the POMS) provides explicit guidance on those who should be *prohibited* from selection as payees (POMS GN 00502.132):

- fugitive felons,
- representatives or health care providers who have committed Social Security fraud, and
- individuals having an unsatisfied felony warrant (or in jurisdictions

that do not define crimes as felonies), a crime punishable by death or imprisonment exceeding 1 year.

The POMS also identifies individuals that generally should be avoided as payees, yet are ultimately eligible to serve as representative payees:

- convicted felons other than those convicted of violations of felonies involving fraudulent action in relation to a benefits applications and misuse of Social Security numbers under the Social Security Act (defined in the Act);
- applicants found guilty of fraud related to the Representative Payee Program;
- applicants with a prior history of misuse;
- people who have been imprisoned for more than 1 year; and
- creditors of the beneficiary.

SSA policy sets priorities for candidates in representative payee preference lists. The POMS provides separate preference lists for payee candidates according to three beneficiary types: minors (children), adult beneficiaries, and beneficiaries with known substance abuse problems (POMS GN 00502105).

When the beneficiary is a *minor child*, select the best payee available from this list of preferred applicants: a natural or adoptive parent with custody; a legal guardian; a natural or adoptive parent without custody, but who shows strong concern; a relative or stepparent with custody; a close friend with custody and provides for the child's needs; a relative or close friend without custody, but who shows strong concern; an authorized social agency or custodial institution; or anyone not listed above who shows strong concern for the child, is qualified, and able to act as payee, and who is willing to do so. For *non-substance abusing adult beneficiaries*, the preferred list is: a spouse, parent or other relative with custody or who shows strong concern; a legal guardian with custody or who shows strong concern; a friend with custody; a public or nonprofit agency or institution; a Federal or State institution; a statutory guardian; a voluntary conservator; a private, for-profit institution with custody and is licensed under State law; anyone not listed above who is qualified and able to act as payee, and who is willing to do so; a friend or relative without custody, but who shows strong concern for the beneficiary's well-being; or an organization that charges a fee for its service.

On the whole, it is clear that SSA policy on payee candidate suitability promotes three seeming principles:

1. Only the most egregious circumstances result in exclusion from appointment to payee.
2. Flexibility in payee selection is critical, especially with respect to beneficiary needs.
3. A less than ideal ("bad") payee is preferable to not having a payee.

Data from the committee's survey provide a useful glimpse of the appointment process. The vast majority of payees actively expressed a willingness to serve their beneficiaries, 96.8 percent (0.5). For the small pool of reluctant payees, their concerns included the responsibility of being a payee, 72.1 percent (5.1), and what SSA required of payees, 66.3 percent (7.0), as well as the time commitment, 38.5 percent (7.4), and the potential impact on the payee's relationship with the beneficiary, 35.6 percent (6.0).

Although it is a very positive sign that nearly all payees serve willingly, a critical dimension of performance is the ability of payees to serve their beneficiaries. Survey results suggested a reasonably large proportion of payees may be in economically unstable circumstances. As discussed in Chapter 2, more than one-half of payees reported individual annual incomes of less than $15,000, although this indicator may not reflect potentially much larger household incomes. The estimated 5-year bankruptcy rate for active payees, 6.4 percent (0.8), appears to be higher than the national rate.[2]

The rate of changing residences for the representative payees is similar to the rate in the general U.S. population. More than 28 percent of the payees had moved residence in the last 2 years, 28.4 percent (1.5); in comparison, nationally in 2003, about 15 percent of the population (over 1 year old) reported moving in the previous year (U.S. Census Bureau, 2004).[3]

During the committee's site visits we also considered the fraction of payees with a history of criminal activity or substance abuse (given that some felons are allowed to be payees under SSA policy). The estimated prevalence of payees who had been convicted of a felony is reasonably low, and the fraction of payees who had served in prison is about the same as the estimated national rate of 2.7 percent (Bureau of Justice Statistics, 2004). A very small percentage of payees admitted that they had undergone drug or alcohol rehabilitation in the last 5 years. Thus, it appears that a small seg-

---

[2]During fiscal year 2006, there were 1,085,209 nonbusiness bankruptcy filings (Administrative Office of the U.S. Courts, 2006). As of April 1, 2005, there were 222,940,420 adults in the United States (U.S. Census Bureau, 2006). Assuming that each nonbusiness filing represents a single individual and that no individual filed twice in a 5-year period, the bankruptcy rate among U.S. adults is 2.4 percent.
[3]Since the survey asked about payees' moving in the past 2 years and the Census Bureau data are for 1 year, it is difficult to assess whether the mobility of payees differs from that of the general population.

ment of the payee population may have characteristics that are associated with reduced ability to effectively serve beneficiaries and with increased risks of program violations or misuse.

Even though the prevalence rate of risk indicators for payees is similar to that of the U.S. adult population, such instability poses a potential conflict of interest and increases the risk of misuse. If people with these characteristics are chosen to be payees, it seems reasonable that additional monitoring would then take place. Such monitoring can occur either from the office that made the appointment or by a designated person in the district or region who is chosen to monitor difficult cases.

Some of the findings in Chapter 5 on misuse deserve repetition. The determination of payee suitability should include the consideration of such factors as felony status, any time spent in jail, previous history of misuse, employment status, credit rating, mobility, whether or not the payee resides with the beneficiary, and whether or not the payee is a creditor for the beneficiary. These factors are currently absent from the exclusion rules in the POMS, but the data collected by the committee showed that most of them are associated with misuse.

Although criminal activity or substance abuse among payees is similar to that of the U.S. adult population, such instability poses a potential conflict of interest and increases the risk of misuse. SSA must be more judicious in establishing the suitability of representative payee candidates for at-risk beneficiary populations.

CONCLUSION The Social Security Administration appoints some payees with characteristics that raise questions about their suitability as payees.

RECOMMENDATION 6.3 The Social Security Administration should screen potential payees (including organizational payees) for suitability on the basis of specified factors associated with misuse, particularly credit history and criminal background.

RECOMMENDATION 6.4 The payees of at-risk beneficiaries should be monitored more frequently and intensively than current protocols provide.

"Individual" Payees with Multiple Beneficiaries

SSA policy allows for an individual to serve up to 14 beneficiaries at a time. The committee's study of possible misuse found several cases in which individuals were payees for numerous beneficiaries and also affiliated with organizations that serve the beneficiaries, possibly fee-for-service.

Such situations constitute conflicts of interest. Although SSA definitions designate these people as "individual" payees, some are clearly operating as organizations or as group homes and some are concurrently defined by the state to be in charge of organizations providing assisted living, board and care, or foster care. In many of these cases, the representative payee is not only the disburser of SSA benefits, but also the provider of services, including shelter and food.

In some cases, the state provides a regulatory schema for monitoring the facility with which the individual payee is affiliated, including rules for the fiscal management of SSA benefits. Some states set the provider's rate for services; in others, the state does not exercise any oversight over the way the beneficiaries were charged. In the latter situation, the payee is free to charge whatever she or he deems appropriate for the services rendered and to deduct the charges from a beneficiary's monthly payments.

The committee's study showed questionable practices regarding the charges levied by payees in such situations: some payees adjusted the monthly room-and-board charges to the total check amount received by a beneficiary so that no additional funds remained (or reserved a nominal weekly cash allotment for the beneficiary's spending money). Such fee policies were used even when different beneficiaries in the same facility received different levels of payment, meaning a different "price" was charged for the same service. Other payees appeared to charge above-market rates in order to ensure the complete exhaustion of monthly checks. In many of these cases, the payees were able to provide detailed records of charges, complete with receipts, so that from an accounting standpoint, the payee was performing according to program policies. However, when a payee is a creditor of a beneficiary, either as a landlord or as a provider of board and care, it is unclear whose interests are being served. When a representative payee is the payee for several people in her or his care (as a landlord or a facility), it is inappropriate for the payee to be considered an *individual* payee.

Selection of an appropriate payee is the most important way of ensuring that a beneficiary's needs are met. Individual payees who are administrators of care organizations or who serve several beneficiaries in some other organized capacity should be treated differently from payees who truly serve beneficiaries as individuals. Such groups are in need of monitoring similar to that provided to organizational payees, and SSA should consider a new category that properly classifies organized care providers.

CONCLUSION The current designation of "individual payee" is too broad a category. The designation mixes payees who serve a single or even a few beneficiaries with payees who operate group homes for up to 14 beneficiaries. Individual payees who are owners or administrators of group homes have an inherent conflict of interest. Payees of this type require special monitoring.

RECOMMENDATION 6.5 The Social Security Administration should develop policies that define and treat as an organizational payee an individual who serves multiple, unrelated beneficiaries and who is also the owner, administrator, or provider of a room-and-board facility.

More thorough investigation and monitoring are needed for such payees than for truly individual payees. The committee suggests that SSA develop performance standards specifically for this type of payee.

RECOMMENDATION 6.6 The Social Security Administration should reevaluate its policies that permit creditors and administrators of facilities to serve as payees.

Dual Roles and Fees for Payees

Several issues may arise when individual payees also have roles as conservators or guardians or hold a power of attorney for their beneficiaries. One issue involves the need for a separate "appointment." For example, under current policy, a person who holds a beneficiary's power of attorney must apply to be the payee if she or he wishes to serve in that role. This application is necessary because SSA does not recognize guardians or court-appointed conservators as a substitute for a representative payee. If they do not apply, SSA may appoint someone else to be payee. Sometimes they are unaware that they need to apply to be a payee at the same time that SSA is unaware of their existence and so is seeking a payee. This situation suggests the need for improvement in communication between state courts and SSA about SSA beneficiaries.

To facilitate communication and information exchange between state courts and the Representative Payee Program, there may be a need to clarify an issue related to the Privacy Act (5 U.S.C. § 552a(b)(11)). The issue is whether the exception to the Privacy Act, which permits disclosure of records maintained by an agency "pursuant to the order of a court of competent jurisdiction," applies to orders of state courts seeking access to records about individuals appearing before such courts in guardianship proceedings and who have served or are serving as representative payees. State guardianship courts, for instance, might wish to access or obtain data on payees (e.g., from the Representative Payee System [RPS]) in order to verify the credentials, integrity, and capabilities of guardian candidates with former experience as representative payees. The committee assumes that state courts are "courts of competent jurisdiction" within the scope of the exception to the Privacy Act. The viability of this Privacy Act exception, and the SSA's acceptance of state court orders as within its scope, is an important element in information exchange and cooperation to assure responsibility and appropriate transparency in the activity and conduct of fiduciaries serving incapacitated individuals, such as guardianship wards and beneficiaries dependent on representative payees.

Another issue regarding dual roles involves fees for serving beneficiaries. The committee's study of misuse revealed a discrepancy between two kinds of payees. There were payees who were "given" fees by the beneficiaries for taking care of them and guardian and conservators who took fees under court authority. Misuse clearly occurs when a beneficiary "gives" a payee a monthly fee (taken from the benefits payments) as compensation for the payee's service as the beneficiary's money manager. However, when a payee takes court-sanctioned fees from SSA beneficiary payments in return for services, the issue of misuse becomes murky.

The issue of "court-approved" fees for guardianship and representative payee services is complicated. There is no legislative or regulatory authority to allow court-sanctioned fees for service as a representative payee. The Social Security Act (42 U.S.C. §§ 405(j)(4(A) & 1383(a)(2)(D)) stipulates that only a few types of payees may draw a fee from Social Security benefits, and an organization must apply and be authorized by SSA in order to collect a fee. To qualify as a fee-for-service payee, an organization must

- be a state or local government agency, or
- be a community-based, nonprofit social service agency, that is bonded and licensed by each state in which it serves as a representative payee (if available), and
- regularly provide payee services to at least five beneficiaries and demonstrate that it is not a creditor of the beneficiary.

Despite this statutory language, SSA told the committee that court-appointed guardians may deduct fees from Social Security benefits (see Appendix C). The basis for this statement is found in the POMS (GN 00602.040):

Guardianship Fee: When an individual is appointed a legal guardian for a competent or incompetent beneficiary, part of the beneficiary's funds may be used for customary guardianship costs (or proceedings) and court-ordered fees, provided

- the guardianship appears to be in the beneficiary's best interests,
- the beneficiary's personal needs are met first, and
- the beneficiary's funds would not be depleted by the guardianship costs.

The federal courts have ruled that relying entirely on program management directives is not sufficient to establish regulatory requirements. In cases brought before the federal courts in recent years they have found that the POMS and the HALLEX[4] are internal documents and do not have the force of law.[5] They may indicate a direction, but they do not have regulatory or statutory authority. In the case of fees for service for representative payees, the acceptability of courts of general jurisdiction to allow fees to be taken out of Social Security dollars by guardians who are also acting as representative payees is in direct opposition to the legislation regarding individuals' ability to accept fees.

CONCLUSION The guardianship and fee-for-service aspects of the program conflict with the congressional intent that individual payees not receive fees from Social Security funds. Although the Social Security Administration Program Operating Manual System provides policy guidance for allowing fees when there is court oversight, this broad allowance of such a practice is not in the best interests of beneficiaries and conflicts with legislative intent.

CONCLUSION Some beneficiaries have Social Security Administration-appointed payees who are different from the people who hold their power of attorney or serve as legal guardian or conservator. This causes potential conflicts, violations of Social Security Administration rules, inefficiencies and inaccuracy in reporting, delays in payee selection, and duplication of effort.

CONCLUSION There is a lack of communication between the Social Security Administration and state courts with regard to beneficiaries who might have both a guardian and a representative payee. This lack of communication has led to misunderstandings as to the authority, or lack thereof, for paying fees for representative payee services.

RECOMMENDATION 6.7 The Social Security Administration should change the Program Operating Manual System to state that when a beneficiary already has an individual with power of attorney, a legal

---

[4]The HALLEX is The Hearings, Appeals and Litigation Law (LEX) manual. Through HALLEX, the Associate Commissioner of Hearings and Appeals conveys guiding principles, procedural guidance, and information to the Office of Hearings and Appeals staff. HALLEX includes policy statements resulting from an Appeals Council en banc meeting under the authority of the Appeals Council Chair.

[5]See *Fowlkes v. Adamec,* 432 F.3rd 90 (2d Cir. 2005) and *Moore v. Apfel,* 216 F.2d 864 (9th Cir. 2000).

guardian, or conservator, there is a preference (with flexibility) for selecting that individual as the beneficiary's representative payee.

RECOMMENDATION 6.8 The Social Security Administration, in consultation with the states, should eliminate inconsistencies between state and federal practices regarding the calculation of payee fees and financial oversight.

The committee suggests that SSA consider several changes in policies regarding payees who are in dual roles:

1. Should individuals who are payees and also guardians or conservators be allowed to calculate and take a fee from funds that include SSA funds?
2. Is there a need for legislation or a regulation that specifies that such fees should only be allowed if there is oversight of this practice by a state court of general jurisdiction and they do not exceed what a fee-for-service organizational payee is allowed to take from benefit funds (as is the case in Illinois)?
3. If a state regulates providers, such as group homes that are also serving as payees, should SSA develop a procedure for determining that the state's requirements for such regulation cover financial oversight of SSA funds so that SSA could substitute state regulation for federal oversight?

TRAINING AND SUPPORT FOR PAYEES

Training

Training is a key factor for developing a cadre of effective representative payees. Data from the committee's survey show that most payees have an adequate understanding of their duties and responsibilities as payees (see Chapter 3). However, field office staff told the committee that many payees do not appear to understand the details of their responsibilities, such as how to keep records, the need to deposit benefits into a separate account, and the need to save money (see Chapter 3). There was a general consensus among the staff that better payee training is needed. As discussed in Chapter 3, the brochure for payees and Internet resources lack specificity and practical knowledge that can train a payee for effective service. One suggestion is to compile best practices for payees and hands-on examples, review them with new payees, and disseminate them to all payees on a regular basis. One field staff member stated:

The pamphlet is good but you need hands-on examples. You need to talk to them [payees] a lot to get them to understand how they might do a good job. Talk to the payees about how to go grocery shopping for the beneficiary. Have two baskets, don't intermingle the contents with your own groceries. Tell them how to work with the bank. Show them how to do a ledger. Give them a ledger book. Send them some in the mail—this also verifies the address. I would give the payee a CD (DVD) to take home and watch. The CD would remind them of their duties and responsibilities.

Refresher training was also mentioned by staff as important but lacking. For instance, it would be useful to remind payees of their obligation to notify SSA of changes in a beneficiary's living arrangements because it may affect the benefit amount. Another strategy is to follow the lead set by state courts that mandate training of conservators and guardians. If SSA policy designates conservators and guardians as payees, then perhaps the trainings could be consolidated.

SSA staff noted that some payees are innumerate, and others have difficulty reading. The existence of payees with limited skills raised the question of whether some minimal skill thresholds should exist for payee candidates. Payees with limited skills may put their beneficiaries at risk for sub par service, but thresholds may eliminate as payees people who would look after the beneficiary beyond the basic stipulations of payee responsibilities, e.g., parents of child beneficiaries. The committee believes that it is possible to develop and implement training for payees with limited skills but who are otherwise suitable payees. The principle would be one of inclusion with regard to payee selection. And it would rely on field office staff being assiduous in their efforts to train and support payees.

In 1996, the Representative Payment Advisory Committee (U.S. Social Security Administration, 1996) also raised related issues that the committee found are still fundamentally unaddressed:

- Lack of a designated person in the SSA field offices to be responsible for and be the contact for representative payee issues.
- Absence of ongoing workshops and seminars for training and supporting payees.
- Lack of ongoing and periodic communication from SSA field staff to payees.
- Absence of materials containing specific examples or workbooks of how payees should carry out their responsibilities.

RECOMMENDATION 6.9 The Social Security Administration should provide comprehensive and formal training for representative payees.

The committee suggests that an enhanced training system might include a more intensive initial training session than exists in current practice and regular refresher sessions. The training might include mandatory oral briefings, enhanced yet simplified hard-copy materials, and examples of best practices, Internet-based training, and a dedicated toll-free hotline for payee assistance.

Support

A component of the Representative Payee Program that seems underdeveloped is an organized system for ongoing support for payees. The committee's survey showed that only a small fraction of payees sought assistance from SSA in relation to their role as payee, and relatively few had used the SSA's Internet site (see Chapter 3). Regardless of whether a payee had previously contacted the SSA office, a significant number of payees were interested in additional SSA assistance.

The SSA website (www.ssa.gov/payee/index.htm) provides the principal source of ongoing support to representative payees. As noted in Chapter 3, only about 1 in 10 have used SSA's payee website, even though nearly two-thirds of representative payees had access to the Internet. This suggests that the effectiveness and perhaps awareness of the site could be improved.

The site visit reports suggested that ongoing training of payees was needed. As mentioned earlier, one idea was to provide DVDs to payees as a visual aid the payee could reference in the privacy and comfort of his/her own home. Another suggestion was to schedule "refresher" training sessions.

RECOMMENDATION 6.10 The Social Security Administration should provide payees access to various types of well-advertised support in their activities. Such support could include: (1) dedicated field staff who can serve as contact persons for payees; (2) toll-free telephone numbers specifically for use by payees to seek assistance from SSA; (3) easily comprehensible brochures containing examples and explanations; (4) enhanced, easy-to-use FAQs and online learning tools; (5) guidance on how to meet accounting and document retention requirements; and (6) online guidance for payees to complete the annual accounting form.

MONITORING AND ACCOUNTABILITY

If the Representative Payee Program is to be effective, performance of payees needs to be monitored, and problems should be addressed expeditiously. Currently, there appears to be very little performance monitoring and accountability for individual payees who serve fewer than 15 beneficiaries and for nonfee organizational payees that serve fewer than 50 beneficiaries. The current methods for monitoring are the accounting form, handling of ad hoc complaints by field staff, and the RPS. The committee found these resources to be inadequate and/or ineffectively targeted, as detailed below.

The Accounting Form

SSA created an annual accounting form for payees after a court case and legislation mandated equal treatment and annual accounting for all payees.[6] The legislation also required the commissioner of the SSA to establish and implement statistically valid procedures for reviewing such reports in order to identify instances in which such persons are not properly using such payments (42 U.S.C. § 1383(c)(i)).

The first monitoring milestone involves flagging the payee's submission of (or failure to submit) an annual accounting form, as required by SSA policy. However, not all payees realize that an annual accounting form is mandatory: the committee's survey showed that 7.1 percent (0.7) of payees did not understand the requirement (see Chapter 3). Although failure to submit the form can be an indicator of improper use or misuse, SSA told the committee that resources are inadequate for fully pursuing and investigating payees who fail to submit annual accounting forms (Appendix B):

> The sheer volume of required annual accountings creates an extremely large administrative burden and prevents SSA from being able to develop an in-depth review of a smaller population of possible higher risk payees.

Although there may be legitimate reasons for some failures to file the form—for instance, a beneficiary's payee may change several times in a year so that no single payee is responsible for a full year—the committee's study of misuse shows that such failure is a useful indicator of potential misuse or improper use.

Once the form is submitted, SSA assesses whether benefit funds are being properly used for the beneficiary. This assessment is largely an exercise in determining whether the entered amounts are consistent and plausible. However, as discussed in Chapter 4, the form can be completed impeccably, yet misuse can still be occurring. And at the other end of the spectrum, a completed form can have illogical or questionable entries, yet further investigation shows that misuse did not occur.

When irregularities or errors do trigger a follow-up with a payee, the

---

[6]The court case was *Jordan v. Schweiker*, 744 F.2d 1397 (10th Cir. 1984); *Jordan v. Bowen*, 808 F.2d 733 (10th Cir. 1987); the legislation is 42 U.S.C. § 405(j)(3).

forms are retrieved and forwarded to local field offices. The SSA staff then contact the payees to investigate further. However, after an annual accounting form is forwarded to a local field office for resolution, the form is not necessarily returned to the designated repository (the Wilkes-Barre Data Operations Center). To the extent that forms are not ultimately returned to the Data Operations Center, a potential indicator of payee misuse—an erroneous or incomplete annual accounting form—is thus not available for further use and analysis. In particular, the RPS will not have information on the completion status of the annual accounting forms that have been questioned and investigated.

Beyond the case-by-case review, SSA does not have an appropriate method for systematically evaluating and validating the material it receives on the annual accounting forms. The forms are now optically scanned at the Wilkes-Barre Data Operations Center for electronic storage, but financial data provided by payees are not entered into a database that can be analyzed for compliance and tabulated to produce an indicator of possible violations. As such, this method does no more than preserve a hard-copy version in electronic form; it does not allow for statistical tabulation or analysis of the reports' contents, as required by legislation.

As part of an examination of lump-sum payments, we requested annual accounting forms for 50 Old Age, Survivors, and Disability Insurance (OASDI) payees and 50 Supplemental Security Income (SSI) payees: 18 percent of the forms for the OASDI payees could not be retrieved, and 28 percent could not be retrieved for the SSI payees. It is not clear if the reports were lost in some local office or were never submitted by the payees. Since many beneficiaries change payees on an almost regular basis (with some payees cycling in and out of the system for the same beneficiaries), it is possible for many months, maybe even years, to pass without a particular payee responding to the annual accounting process.

Another issue associated with the annual accounting form is perceived paperwork burden on payees. Some field office staff expressed concern about the burden placed on payees. However, the committee's survey showed that 80.9 percent (1.2) of payees said the form was easy to fill out, and only 10.6 percent (0.9) reported some difficulty completing it.

On average, representative payees spent 14.8 (0.4) minutes completing the annual form. The vast majority of payees, 82.4 percent (1.2), spent less than 1 hour and only 12.2 percent (1.0) spent 1-2 hours to complete the form. The apparent ease of completing the annual accounting form, combined with the fact that one-third of payees did not keep financial records (and many more did not organize their records), indicates that the current annual accounting form is not an effective or trustworthy tool for monitoring.

Although the accounting form is thought to be a psychological deterrent (at best) for ensuring that funds are spent on beneficiary needs, in its present form it does not provide sufficient information to determine if benefit funds are used appropriately. The committee believes the form can provide much better needed information and proposes a redesign for SSA consideration: See Chapter 5 for discussion and Appendixes E and F. Another revision to the process would be online filing (see Chapter 5).

During its site visits, the committee encountered a larger problem of the lack of effective training and tools for payees to track expenses. In fact, there are no guidelines as to which receipts payees must keep to verify appropriate use of beneficiaries' funds. More generally, there is no standardized bookkeeping or accounting method that is required of payees.

CONCLUSION The statutory provision (42 U.S.C. § 405(j)(3)(A) (OASDI); 42 U.S.C. § 1383(a)(2)(C) (SSI)) that requires the Commissioner of the Social Security Administration to establish and implement statistically valid procedures for reviewing the annual accounting forms creates a concomitant obligation to provide information for understanding and monitoring the performance of representative payees. This obligation is not being fulfilled.

CONCLUSION The filing of annual accounting forms (or the failure to file them) is not reconciled with any other administrative record so that a failure to file would bar a payee from continuing to serve in such a capacity.

CONCLUSION It is too easy for representative payees to learn that if they just fill out the accounting form with some plausible, but possibly inaccurate information, they will have complied with the program's reporting requirement and that there will be no follow-up or other consequences. Essentially, the current monitoring process is an "empty threat" that can easily be subverted and is an expensive administrative tool that does not yield the sort of data that are necessary to uncover misuse.

CONCLUSION The Social Security Administration does not have a method for systematically evaluating and validating the material it receives on the annual accounting forms. The data on the accounting form are not retrievable for statistical analysis and therefore, empirically based policies and regulations cannot be formulated. In addition, the Social Security Administration's legislative obligation to statistically tabulate the annual accounting form remains unfulfilled.

RECOMMENDATION 6.11 The Social Security Administration should reengineer the annual accounting form to ensure the usefulness of the data and their transferability into the Representative Payee System and other Social Security Administration information systems.

RECOMMENDATION 6.12 The Social Security Administration should store data from the annual accounting forms in an electronic database suitable for analysis.

RECOMMENDATION 6.13 The Social Security Administration should provide the option for payees to complete the annual accounting form online.

Review of Ad Hoc Complaints

A second monitoring strategy in the program is the review of ad hoc complaints of potential misuse by field office staff. Typically, misuse complaints are lodged against a payee by the beneficiary or nonpayee relatives or friends of the beneficiary. A claims representative is charged with investigating such complaints. The initial inquiry typically leads to a complex array of facts, events, assertions, and interactions among the accuser, the payee, and the beneficiary (see below). However, as discussed in Chapter 4, the claims representative has little incentive to pursue a formal investigation. The amount of time, analysis, and paperwork that is required to establish misuse, coupled with a malleable definition of misuse, leads claims representatives to avoid the difficult task of documenting misuse. The usual approach is for the claims representative to determine whether or not the beneficiary should have a new payee, and if so, to "deselect" the current payee and identify and appoint a new one. Although most cases of formal misuse do arise from complaints, the vast majority of complaints do not result in a formal finding of misuse. Rather, they usually result in the replacement of a payee (see Chapter 4). Thus, potential and likely misuers are not held accountable, and they can—and frequently are—appointed again as a payee to the original beneficiary or to another beneficiary.

We illustrate this scenario with a situation we found many times in our site visits: custody changes for minors with parents who do not live together. Claims representatives reported that they devote a substantial amount of their time to sorting out child custody issues. Each time a custody change occurs, they must verify the child's residence through administrative data, such as school records. Shared custody situations lend themselves to conflicts between parents (or other custody sharers). One claims representative reported that a parent continued receiving the child's benefit checks even though the child had moved to the other parent's residence. Another reported finding that in checking a parent's claim for a resident child, an approved claim had already been filed by the other parent who was receiving the child's checks as the payee which constitutes misuse. It is clear that shared custodial beneficiary situations may facilitate or mask misuse and require continuous and careful monitoring by SSA staff.

The difficulty of defining misuse in specific complaint situations was reported repeatedly in the site visits. Explicit definitions and guidelines appear in the Social Security Act and the POMS (see discussion in Chapter 5). In response to a committee question, SSA stated:

> The (SSA) technicians need to use information contained in the POMS and use their own judgment and expertise to resolve the problems or issues presented to them in each particular case. It would be impossible to cover by written instruction every conceivable situation the field will run into given the size and diversity of this population.

Site-visit reports note that as a result of this difficulty, field staff tend to deselect and replace payees instead of a formal finding of misuse, and the standards used to determine misuse are inconsistent across local offices and staff. For instance, some claims representatives included as misuse instances in which a payee did not know whether or not the beneficiary's needs were being met, regardless of how the funds were being spent. Others included as misuse the practice of a payee taking a small fee from the beneficiary payments, although SSA allows this in specific circumstances (e.g., court approved fees to conservators), or a failure by a payee to respond to contact attempts by SSA.

CONCLUSION Factors such as lack of incentives for staff to investigate misuse, perceived vagueness in the definition of misuse, and the complexity of interpersonal relationships between beneficiaries and their payees often lead claims representatives to find a more suitable payee rather than to formally determine misuse.

CONCLUSION Frequently changing custodial arrangements for beneficiaries who are children involve complicated situations that may facilitate payee misuse.

RECOMMENDATION 6.14 The Social Security Administration should establish mandatory protocols for payee replacement when misuse is suspected.[7] When misuse or suspected misuse is the reason for a change of payee, staff should provide full documentation.

The use of phrases such as "more suitable payee found" should not be allowed as formal documentation.

---

[7]Consistent, of course, with the OBRA Amendments, 42 U.S.C. § 405(j)(2)(E), 1631(a)(2)(B)(x)-(xii), which provides payees and "any individual who is dissatisfied with a determination by the Commissioner of Social Security to certify payment of such individual's benefit to a representative payee . . . or with the designation of a particular person to serve as representative payee" with the right to [an ALJ hearing and to judicial review of the Commissioner's final decision]." Such hearings are apparently rarely requested, however.

The Representative Payee System

The RPS is a database system used to enter and maintain information about representative payees and the beneficiaries they serve. The RPS is mandated by statute and requires SSA to establish and maintain a centralized file, readily retrievable by SSA offices, which contains the names of payees who have had their status revoked by reason of misuse of funds or because of a program violation. The RPS is thus a critical part of the system to administer the payee program, and it is heavily used by SSA offices to document payee relationships and evaluate potential payees for service. However, our attempts to use the RPS as a research tool and information from field office staff during our site visits indicate that the system could be greatly improved to more effectively facilitate SSA's mission, both for monitoring individual payees and for agency-level studies of the program. We note that SSA is in the very early planning stages for a revision to the RPS. Thus, our discussion in this section is based on our study of the current system.

A critical problem with the RPS is that it does not contain entries for all payees: SSA estimates that several thousand active payees are not included in the system (Appendix B). In some, but not all cases, the omissions are due to a systematic barrier. In particular, payees without a Social Security number (SSN) (e.g., undocumented alien parents, foreign nationals) cannot be registered in the system. This situation provides an opening for field staff to avoid entering other cases because of the difficulties encountered with the RPS, something that was observed during site visits. SSA acknowledged this problem (Appendix B):

> There are a significant number of cases missing from the RPS. Bypassing the payees is of great concern because it prevents SSA from affording protections as designed. Because some rep payee applications cannot be taken in (undocumented alien parent payee without an SSN), we had to have processes in place so these cases could be processed outside of the RPS, a situation which also allows abuses to occur.

A second significant problem is the difficulty of using the RPS to record information on payees at time of application. The process for doing so is laborious, in part due to the outdated interface design for the RPS. Essentially, users are required to navigate through irrelevant screens on their way to a target screen and then to back out of screens to pursue a new query; this is cumbersome and illogical to users. Moreover, the office workload does not allow much time for each visit with a potential payee so inefficiencies in the system discourage complete and detailed data entry.

Because specific RPS conventions for entering data fail to facilitate efficient and accurate data entry, field staff are often uncertain about what additional information to provide and the amount of detail to enter into the system. For example, the RPS does not require entry for many important variables even though policy dictates that they be recorded. If the claims representative does not do forced entry of the variables—using a mutually exclusive and exhaustive set of response options—there will be missing data that can hamper later investigations about a payee's service.

There are also many instances in which critical data are entered as free text, which make it difficult to achieve consistency across offices and staff and relatively easy to forget to record important information or omit important details. For example, as discussed above and in Chapter 4, "more suitable payee" is frequently entered as the reason for terminating a relationship and choosing a new payee. This phrase communicates very little to the next staff person who may need to evaluate the terminated payee. Since a lot is known about payees and the reason for their terminations, it would be possible to create forced choices—a checklist—for many data elements that would provide more meaningful and simpler response options for users. The current situation enables "office shopping" by some payees because problematic histories are difficult to uncover. Furthermore, although the RPS has automatic edit checks for some entries, without forced entry and closed-ended responses for critical variables, effective data checks cannot be implemented.

A third problem related to the outmoded design of the RPS is that the system is not being effectively used by all offices, and some of its features are not used by any of the offices. For example, some field staff do not enter enough notes to describe situations observed during interactions with beneficiaries and payees. Such notes are not enforced by the software, yet they could alert staff to repeat issues and complaints or provide valuable background for new payee selections or a change in payee. It also appears that staff training on the RPS would be valuable: for example, some staff believe that they must reenter a name and SSN, although the system does not require this.

Updating the RPS is a fourth important problem. Updates are entered on an ad hoc basis by field staff in response to notification from a payee or beneficiary that circumstances have changed. When the committee attempted to use the RPS to understand the circumstances surrounding misuse cases, current information about a payee's employment, financial circumstances, or family living arrangements was repeatedly absent. Although payees are responsible for reporting updates, there is no formal system to ensure that this occurs. This is another factor that reduces the currency and quality of payee and beneficiary data. Ideally, the RPS should also be updated with information on the annual accounting forms to establish which payees submit annual accounting forms and which do not. This is a critical linkage because the failure to submit the form can be used to monitor payee performance and to investigate potential violations and misuse.

CONCLUSION The Representative Payee System is a badly flawed tool for case-by-case field use to evaluate prospective representative payees and to investigate problems with payees. Office-to-office autonomy regarding procedures for making entries into the Representative Payee System and a cumbersome and inefficient interface create an environment that encourages inconsistencies in the amount and quality of information available in the database. In addition, data quality concerns and incompleteness compromise the potential for the Representative Payee System to be used for research and analysis with aggregate data, such as summarizing characteristics of the payee population, investigating factors associated with misuse, and drawing samples for monitoring payees.

RECOMMENDATION 6.15 The Social Security Administration should redesign the Representative Payee System.

The committee suggests that SSA consider the following changes to the RPS:

1. inclusion of all payees into the system;
2. creation of data elements in the system with respect to a payee who is identified as a potential or suspected misuser;
3. addition of data elements in the system for various types of violations by payees;
4. addition of data elements in the system for relevant results of investigations by the Office of the Inspector General;
5. inclusion of the Employer Identification Numbers of all organizational payees in the system;
6. addition of a lump-sum indicator and amount to enable the local field office to better monitor how such money is spent for a specific beneficiary;
7. easy access and use by all field office staff;
8. streamlined linkage to annual accounting form data; and
9. an easy-to-use interface that has undergone usability testing.

The committee suggests that SSA require entry of important data elements—standardized values that ensure consistency of responses across offices and support institutional analysis of the full population or special populations of payees. A new RPS should undergo usability testing to ensure that it effectively supports office staff in entering and updating the system. These improvements could be logical considerations under SSA's currently planned revision of the RPS.

RECOMMENDATION 6.16 The Social Security Administration should implement a process that regularly updates information in the Representative Payee System, both by field office staff and through the annual accounting form. The Social Security Administration should also implement a quality control program that periodically checks the integrity of the information in the Representative Payee System.

TERMINATIONS AND TRANSITIONS

There are many reasons that a payee might cease to serve a particular beneficiary, including:

1. The beneficiary requests a change in payee.
2. The payee requests termination.
3. Payee misuse is determined.
4. SSA identifies a more suitable payee.
5. The beneficiary no longer needs a payee (e.g., an emancipated minor, a minor aging into adulthood, or a beneficiary's recovery from a period of incapacitation).
6. There is a change in custody for the beneficiary.
7. The representative payee or beneficiary dies.

When there is a transition from one payee to another, the selection of a new payee is required to be conducted with the same level of scrutiny as an initial selection of a payee (POMS GN 00504.100). Also, when there is a transition, POMS (GN 00605.360) states that a final accounting of beneficiary funds must be conducted.

In addition to voluntary transitions from one payee to another, SSA has the authority to terminate any payee who is not performing according to the standards outlined in the representative payee brochure. Changes in payees, for whatever reason, are not a common occurrence. In the committee's survey, more than three-fourths of payees, 79.9 percent (1.2), had never experienced a termination of their tenure. An estimated 15.0 percent (1.0) of payees reported it happened once, and 3.3 percent (0.6) reported two terminations. Only 1.8 percent (0.9) reported being terminated more than twice.

The committee investigated the reasons for the terminations reported in our survey. On the whole, the reasons reflected understandable, logical circumstances for ending the payee appointment. More than one-half, 55.8 percent (2.6), occurred because of a change in beneficiary eligibility (the beneficiary died or otherwise became ineligible for benefits). Among payees associated with an organization, 60.4 percent (6.5) were terminated because the beneficiary had been discharged or was no longer receiving services.

A smaller fraction of payees initiated their own termination because they did not wish to continue being a payee, 10.4 percent (1.3), or because they could not meet the beneficiary's needs, 6.5 percent (1.1). About the same percentage of payees were terminated at the beneficiary's request, 10.9 percent (1.7). In a small number of cases for current payees who had previously been terminated, SSA initiated a termination, 3.6 percent (0.9).

As discussed above, the committee found that when an ad hoc complaint of misuse was filed, a formal investigation was regularly bypassed and replaced with deselection of the payee under suspicion and appointment of a more suitable one. To explore the prevalence and circumstances surrounding this practice, the committee investigated reasons that payees were terminated, using a 1-percent sample of just under 143,000 records drawn randomly from the RPS. We note again the 58 percent of the terminations noted in the RPS carry the notation "more suitable payee found." The next most common reason for termination was "other," in 24 percent of cases. Unfortunately, neither of these reasons provides any useful interpretive information that can be used in the specific case or for statistical analysis. The only other common reason cited was "benefit ceased," 6.5 percent.

Payees who are terminated due to suspicions of misuse but without a formal investigation and finding, remain available for appointment to another beneficiary or to continue serving as the payee to other beneficiaries. We also note that the committee found that when misuse was suspected and a more suitable payee found, a final accounting was not always conducted as required (POMS GN 00605.360).

CONCLUSION Whenever suspected cases of misuse are not subjected to a formal investigation but handled by use of the phrase "more suitable payee found," potential misusers are not held accountable for their actions; this approach may actually promote reentry to payee status (for some other beneficiary) and consequently future misuse.

CONCLUSION Lack of the required final accounting for terminations may cover up misuse, especially in cases in which a "more suitable payee" was found.

RECOMMENDATION 6.17 The Social Security Administration should revise the current regulations that require a final accounting whenever a payee is terminated to ensure, so far as practicable, that all funds are accounted for.

STATE-RELATED ISSUES

State policies substantially influence the administration of the Representative Payee Program across the United States. The variations in state-specific program implementation are due to such policies as:

- State courts allow fees to be paid to guardians who also administer SSA funds as representative payees.
- State courts have oversight over guardians and conservators who also are payees and impose bonding, training, and accountability requirements that are generally greater than SSA has for those who serve only as payees.
- Mandatory provisions for providers of assisted living or boarding homes to be the representative payee for those who reside in those homes.
- Former residents of state mental health institutions are housed in boarding houses where the owner/administrator of the facility is also acting as the payee.

Over the past few decades there has been an increase in the monitoring and oversight that state courts exercise over guardianships and conservatorships, through both legislation and court rules. On an individual level, the amount of increased scrutiny depends on the specific court and its resources.

State-appointed guardians who are also payees, in general, must report all of their financial activity to the court. Although there may be greater oversight of payees by state courts than by SSA, SSA still has responsibility to beneficiaries to ensure that there is no misuse of Social Security benefits. Nevertheless, the committee's study suggested that payees were deferring to state courts for guidance and authority on how to spend and report financial information for their beneficiaries, including stipulations on how the guardians were to be paid from beneficiaries' funds that included Social Security benefits (see Chapter 5). Fees have been authorized on the basis of the time spent on guardianship services, as well as time spent filing lawsuits, filing taxes, and other services. Guardians reported that as long as there was "approval" by the state court that would be acceptable to SSA. Yet there is no place on the annual accounting form (or any other required payee paperwork) to show what has been filed with the state court or the court's acceptance of the filing. There is currently no "deemed status" between SSA and state courts that would lead SSA to accept the reports filed with a court in lieu of the SSA annual accounting form.

CONCLUSION Funds from both the Old Age, Survivors, and Disability Insurance (OASDI) Program and from the Supplement Security Income (SSI) Program are used to pay fees for representative payee services without regard for legislative limitations because of the way in which the Social Security Administration defers to state court oversight of guardianship and conservatorship financial reporting.

State legislatures, special court task forces, and state courts themselves have been under pressure to improve their guardianship and conservatorship oversight. Requirements for training and monitoring of guardians have increased. Some states require guardians to be certified (a national certification has been developed); others require guardians to complete a course of study. States and individual courts may monitor both the financial reports and the people on a regular basis.

CONCLUSION State court guardianship and conservatorship programs operate totally independently from the Social Security Administration Representative Payee Program even though the program requires any beneficiary who has a guardian or conservator to also have a payee appointed by the Social Security Administration. There is no coordination between the Social Security Administration and state courts for the training of guardians, conservators, and payees or regarding filing annual reports.

RECOMMENDATION 6.18 The Social Security Administration should track state laws that require conservators or legal guardians of beneficiaries who need representative payees to undergo court monitoring and mandated training. In such states, the Social Security Administration should give preference to designating the guardians or conservators as the payees and seek to integrate or coordinate its payee training materials with the state-mandated training.

As noted above, although a guardian may also be a representative payee, there is no required communication between the state courts and SSA. Such communication is likely to improve mutual understanding and coordination between the states and the payee program: see earlier discussion, "Dual Roles and Fees for Payees," and Recommendation 6.7.

There is one state that required its providers of assisted living or board-and-care to become the representative payee for residents of the home whether the beneficiary is in the home for a short time, as in an emergency placement, or on a long-term basis. The provider, owner, or administrator then applies to SSA to become the payee as a condition of having the resident placed in the home. The committee is concerned as to whether a state should be conducting the "selection of payee" rather than an SSA claims representative. The payee in these cases is also a creditor and the provider of all services. This situation is one that the committee recommends be reconsidered by SSA.

A second type of state variation is the availability of state funds for additional services for certain classes of beneficiaries based on their medical or social needs. Although some states have very specific guidelines for providers of board and care (in assisted living facilities, group homes, etc.) that afford financial protections to beneficiaries, others do not. In some states the availability of additional resources from Medicaid or state funds to providers of board-and-care homes does not appear to lead to extra monitoring to ensure compliance with proper financial management. Rather, in some cases there is a cash amount provided to the facilities owner or administrator for services to a particular or multiple beneficiaries without any indication as to how or to whom the funds were to be accounted. If such funds are commingled with SSA funds, there are opportunities for misuse or inadequate fiscal accounting.

Another difference identified in state policy is the manner in which states or local jurisdictions allow for the development of board-and-care facilities. In one state it appeared that the state did not regulate such facilities and the services that are provided are entirely up to the providers. The committee also observed the use of additional community services for large numbers of beneficiaries who were taken to day care facilities for several hours each day where they were provided meals, at the same time that the providers claimed to be providing "board-and-care." Differences in state policy, guardianship, social services programs, etc. all may contribute to differences in the states' ability to monitor actual and potential misuse of SSA (and other) funds. The role of the state and that of SSA in ensuring that beneficiaries receive the services that their SSA benefits are supposed to cover needs to be further studied. Policies and procedures need to be put in place to prevent opportunities for misuses or inadequate fiscal accounting.

State policies—such as the licensure of board-and-care facilities, the availability of state funds for additional services, and the ways that states monitor those providers that receive additional financial support—are essential to the accountability of SSA funds. That accountability in turn is essential to ensure that benefit funds meet the appropriate needs of the beneficiaries. An aggregation of all state laws, regulations, policies, and practices that may affect the use of beneficiaries' funds would enable the development and sharing of best practices between the states and SSA.

RECOMMENDATION 6.19 The Social Security Administration should begin an outreach program with state agencies to compile the laws and practices and study the differences in various states' regulation of assisted living, foster care, and other group homes.

The committee suggests that SSA learn about promising techniques and approaches to monitoring and regulating state activities that affect the Representative Payee Program; ensure that states do not unduly dictate the designation of the owners or administrators of group homes as representative payees; and ensure that organizational payees do not receive federal benefits for state-remunerated activities.

VARIATION OF LOCAL OFFICE MANAGEMENT

Throughout this chapter and report we note information gathered from our visits to local SSA offices. As we note above, we were impressed with the dedication and energy that staff exhibited as they tried their best to make the payee program work for beneficiaries. We also note, however, that local management plays a vital role in how the program operates, and we observed significant differences in how local offices were managed. The variations were significant: different practices, different attention to the rules, different interpretation of policies, and most importantly, different allocation of resources to the payee program. The committee did not plan to formally evaluate office differences and the affect of local management on the payee program. But we would be remiss if we did not mention these differences. It is possible that the Representative Payee Program is affected less by formal written policies and procedures and more by leadership, management, systems, and staff differences.

References

AARP
2005   *Money Management Program*. Washington, DC: AARP.
Administrative Office of the U.S. Courts
2006   Bankruptcy Statistics. Table F-2. Washington, DC: Administrative Office of the U.S. Courts. Available: http://www.uscourts.gov/bnkrpctystats/statistics.htm#quarterly [June 2007].
Bureau of Justice Statistics
2004   *Criminal Offender Statistics*. Washington, DC: U.S. Department of Justice. Available: http://www.ojp.usdoj.gov./bjs/crimoff.htm [June 2007].
Elbogen, E.B., C. Soriano, R.Van Dorn, M.S. Swartz, and J.W. Swanson
2005a   Consumer views of representative payee use of disability funds to leverage treatment adherence. *Psychiatric Services* 56:45-49.
Elbogen, E.B., J.W. Swanson, M.S. Swartz, and R.Van Dorn
2005b   Family representative payeeship and violence risk in severe mental illness. *Law and Human Behavior* 29(5):563-574.
Teaster, P.B.
2003   When the state takes over a life: The public guardian as public administrator. *Public Administration Review* 63(4):396-404.
U.S. Census Bureau
2003   *American Community Survey Profile*. Washington, DC: U.S. Census Bureau. Available: http://www.census.gov/acs/www/Products/Profiles/Single/ACS/Narrative/010/NPO1000US.htm [June 2007].
2004   *Current Population Survey, Annual Social and Economic Supplement*. Washington, DC: U.S. Census Bureau. Available: http://pubdb3.census.gov/macro/032005/perinc/toc.htm [June 2007].
U.S. Social Security Administration
1996   *Final Report*. Representative Payment Advisory Committee. Baltimore, MD: U.S. Social Security Administration.

2002   *Pilot Strategy for the Use of Stored Value Cards in the Social Security Administra-*
        tion's Representative Payment Program. Issue Paper A-13-02-22096. Office of the
        Inspector General. Baltimore, MD: U.S. Social Security Administration.

2005   *Nation-wide Review of Individual Representative Payees for the Social Security*
        Administration. Evaluation Report A-13-05-25006. Office of the Inspector General.
        Baltimore, MD: U.S. Social Security Administration.

2006   *A Guide For Representative Payees.* SSA Publication No. 05-10076. Baltimore, MD:
        U.S. Social Security Administration.

Appendix A

Westat Survey Methodology and Survey Questionnaires

This appendix is available online at www.nap.edu.

Appendix B

Program Questions to and Answers from the Social Security Administration

T his appendix presents the committee's questions to the responses from the Social Security Administration (SSA) on the following topics: direct deposit, selection and training of representative payees, misuse, the accounting form, the Representative Payee System (RPS), and the Program Operations Manual System (POMS). The questions arose from the committee's visits and interviews with local field offices. The questions below were formally submitted to SSA in writing, and the agency's formal written responses are reproduced verbatim.

Direct Deposit

1. **Does SSA have any yearly statistics on the use of direct deposit for beneficiaries with a payee?**

The latest data readily available are from 2002. Direct-deposit use by those with and without a representative payee is reflected by the following:

Retirement and Survivors payments without a payee	84%
Retirement and Survivors payments with a payee	61%
Disability payments without a payee	73%
Disability payments with a payee	58%
Supplemental Security Income payments without a payee	56%
Supplemental Security Income payments with a payee	41%

Current data reflect a modest overall increase in the use of direct deposit for payments administered by SSA from 76.4 percent in 2002 to 79.8 percent in 2006.

The agency supports the use of direct deposit and looks forward to any ideas that would increase its usage within the environment of limited operational resources.

2. Does SSA encourage direct deposit whenever it is feasible? If so, how does it encourage it?

Yes, during the application process, we advise applicants of the benefits of using direct deposit and encourage them to use it. Also, periodic stuffers are included with paper checks encouraging individuals to switch to direct deposit. During contacts with our rep payees, such as during a site visit, we encourage them to use direct deposit. However, some of our volume payees have expressed a reluctance to use direct deposit because they do not want to wait for their monthly bank statements to ensure payments were actually received (in case of a change of payee).

SSA supports Treasury's comprehensive direct-deposit marketing campaign called "Go Direct." Go Direct is a national campaign to motivate more Americans to select direct deposit for their Social Security and other federal benefit payments. Information about Go Direct can be found online at http://www.godirect.org/about_faq.cfm.

Again, we would welcome any new ideas you may have to encourage payees to sign up for direct deposit.

Payee Selection

3. Has SSA considered assembling a nationwide pool of professionals to serve as a corps of payees ready to serve in emergency of difficult cases?

Over the years, SSA has conducted outreach efforts to recruit new payees, particularly in metropolitan areas that generally have a larger population of disabled persons with mental and addiction issues. During our outreach efforts, we explained that while the majority of payees do it on a voluntary basis, some organizational payees may qualify to collect a payment for their payee services.

Section 205(j)(3)(G) of the Social Security Act <u>requires</u> field offices (FOs) to keep a list of payee sources located in the local service area and community. In POMS GN 00502.100D, FOs are encouraged to develop ongoing, cooperative relationships with community social service providers who can often provide payee contacts.

In spite of our efforts, some areas of the country have trouble finding qualified payees to serve our more challenging clientele. We welcome any suggestions you may have to recruit new qualified payee sources, provided it does not present major resource burdens on our field office components.

4. Could such payees be paid on a fee-for-service basis?

Yes, but only if they meet specific qualifications as provided by current SSA law. Sections 205(j)(4)(A) and 1631(a)(2)(D) provide authorization that permits certain types of organizational payees to collect fees from the beneficiaries for the payee services they perform. An organization must apply and be authorized by SSA in order to collect a fee. To qualify as a fee-for-service payee the organization must be

- a state or local government agency, or
- a community based, nonprofit social service agency, that is bonded **and** licensed by each state in which it serves as a payee (if available), and
- regularly provide payee services to at least five beneficiaries and demonstrate that it is not a creditor of the beneficiary.

Any criteria beyond these would require legislation.

5. Cases involving beneficiaries who are homeless and/or have substance-abuse problems can be very difficult for most individual payees. Has SSA considered requiring that they be served by professional, fee-for-service payees?

In sections 205(j)(2)(C)(v) and 1631(a)(2)(B)(vii), Congress expressed its preference that beneficiaries with substance abuse problems be served by organizational payees. In general, they have access to greater resources and community contacts and are less susceptible to coercion or intimidation from the beneficiary.

However, Congress allowed SSA discretion in making payee appointments for substance abusers. SSA would not support giving up that discretion because we believe that each payee appointment should be evaluated based on the unique facts of the case. Our primary concern is the well being and best interests of the beneficiary and, while a community based organization may be best for most of these clients, there may be better options for other beneficiaries.

SSA has no specific preference for organizational payees for the homeless and Congress has been silent about the need for a preference for that group.

6. **In such cases (and those in question 2), could the fees be paid from funds not deducted from the beneficiary's payment?**

Without legislation, SSA has no authority to allocate funds from any other source for this purpose. Sections 205(j)(4) and 1631(a)(2)(D) of the Social Security Act provide for qualified organizations to collect a fee from the beneficiary. Organizations can and do seek funding from other sources, generally in the form of grants.

7. **What are the bonding requirements for commercial payees? Has SSA considered relaxing these requirements to make it more feasible for small commercial payees to take difficult cases that are unsuitable for individual payees?**

For clarification, we understand the use of the term "commercial payee" to mean all organizational payees. As we stated in item 2 above, only non-governmental fee-for-service payees are required by law to carry a bond and that bond must be of a sufficient amount to repay funds potentially lost due to a misuse event. No other type of payee, regardless of size, is required to carry a bond, although many do. We would be interested in learning whether some of our payees were interested in becoming a fee-for-service payee but were dissuaded due to the bonding requirement. We welcome any possible recommendations you may have in this regard.

8. **Many of the most difficult selection and reassignment cases involve custody disputes and other heated family disputes. These issues take up lots of staff time, most of whom are not trained to handle these kinds of matters. Could SSA find a way to make social workers or mediators available to offices in these situations?**

In these situations, our mission is to select the one person who will best serve the interests of the child. Although these can be difficult situations, the SSA decision maker can generally determine which parent/family member provides the greater care for the child and who is generally responsible for making purchases and satisfying the child's needs.

Additionally, SSA interviewers have lists of a variety of referral services available to help beneficiaries and their families with a multitude of issues, including various mental health and counseling services. Our technicians are trained to offer names and numbers of the referral services upon request or when they feel it may be beneficial to our clients. We would be interested in suggestions on how SSA can make better use of available community resources.

Payee Training

9. We were told that SSA's written training materials containing guidelines and instructions for being a payee were helpful but were not always available in all field offices? Does SSA believe this is a problem or aware of this issue?

We do not believe this to be an issue. All training materials are available to all field offices through our normal supply channels. In addition, all materials are also available online. The payee application includes reporting responsibilities and is given to the payee at the time of application. "A Guide for Representative Payees" is mailed to the selected payee once SSA makes the decision as to who will best serve the interest of the beneficiary.

10. Are there any written standards for volunteer groups that wish to take on payee responsibilities?

Yes, we have produced a booklet entitled "Guide for Organizational Representative Payees" which provides guidelines and suggestions to assist organizations in understanding the representative payment program. Also the pamphlet entitled "A Guide for Representative Payees" is mailed to all those selected as payee and can be requested by any group who wishes additional information. We also provide a fact sheet for organizations needing information on how to be authorized to serve as a fee-for-service payee.

In addition, we have developed a training package for our field offices to use in educating volunteer groups and organizations that wish to be representative payees. The package, which was updated this year, includes a video, a lesson plan and the "Guide for Organizational Representative Payees" booklet. The package may also be given to organizational payees for in-house training on payee duties and responsibilities.

All these materials are available to the public through the "Social Security Online" website, or by contacting one of our field offices.

Misuse

11. Lack of incentive on the part of office staff to develop misuse cases seems to be a problem. We were told that it takes an average of 10 hours to document a misuse case and that employees receive no "points" for this work. Since "points" determine employee bonuses, employees lack the incentive to do this kind of work. Is that comment regarding "points" true, and what is SSA's response to this question regarding incentives?

While we have heard this complaint from various technicians for a number of years, we have been assured by the Central Office Budget personnel that, as long as work on these cases is properly recorded in the Time and Attendance System, the field office receives appropriate work credit. In addition, the Agency has a new initiative referred to as the Social Security Unified Measurement System (SUMS) which, when fully implemented, will more accurately track workload credit. We hope this will resolve the operational concern.

12. Incentives aside, there is also a widespread perception that the SSA Inspector General will not pursue a misuse case below a certain dollar amount (we were told minima ranging from $10,000 to $20,000). Because of this, office staff will not follow up on smaller cases. Is this true, and is SSA aware that the staff factor this into their decisions in selecting a new representative payee and do not document cases of misuse?

First, we want to emphasize that two provisions of the Social Security Protection Act of 2004 expand OIG's ability to take action against representative payees. Section 111 gives us the authority to impose civil monetary penalties against misusers. Section 201 gives us the ability to impose civil monetary penalties on representative payees who knowingly give misinformation. Therefore, we can take action even if the Department of Justice is unwilling to prosecute a case.

Whenever SSA receives an allegation of misuse, it must determine if misuse actually occurred. If it has, SSA is responsible for locating a new payee for the beneficiary, notifying and pursuing recovery of the misused funds from the prior payee. SSA's actions prevent any continued abuse by the rep payee and are completely separate from any action taken by the OIG. Only after SSA has completed its work on the misuse case is it referred to the OIG for consideration of criminal prosecution. The OIG does not prosecute the case; rather, they review the information developed by the SSA FO and present appropriate cases to the U.S. Attorney who then makes the decision whether or not to prosecute. While the OIG has assured us they do not have a dollar threshold in attempting to present the case to the U.S. Attorney, we believe that it is unlikely that the U.S. Attorney could be persuaded to prosecute a case unless it involves a significant amount of money or contains some other compelling reason to do so. However, the U.S. Attorney's decision to prosecute has no bearing on SSA's responsibility to complete its actions on such a case.

If SSA makes a misuse determination, staff records the misuse information in the RPS. Thereafter, the misuse information will be displayed anytime a future application is filed by the payee or a representative payee query is requested. (More information about this is included in the answer to question 18, under RPS section, below.)

With the passage of the Social Security Protection Act of 2004, we now have additional tools to make beneficiaries whole in cases of misuse and can use our overpayment procedures to recover funds from misuser payees. These administrative actions can be imposed whether or not the case is taken by the U.S. Attorney for criminal prosecution.

Accounting Form

13. What is SSA's view as to its current legal obligations to conduct annual accounting?

SSA is required by law [sections 205(j)(3)(A) and 1631(2)(C) of the Act] to conduct annual accounting for all representative payees except for those payees who are participants in the triennial onsite review program. While we believe the law was well-intended, to afford protections to all of our clients, it requires that SSA conduct annual accountings on nearly 7 million beneficiaries each year, regardless of the relationship, custody arrangements, benefit amount or any other case characteristic. The sheer volume of required annual accountings creates an extremely large administrative burden and prevents SSA from being able to develop an in-depth review of a smaller population of possible higher risk payees.

We look forward to seeing the analysis and conclusions from your study in this regard and any recommendations you may offer that will streamline our current accounting process while protecting our beneficiaries.

14. What leeway does SSA feel it has to modify the form, periodicity, or extent of the accounting?

To be compliant with the law, the payee reporting must be conducted annually and provide sufficient information to allow SSA to judge whether expenditure of Social Security funds is in the best interest of the beneficiary. The existing payee report forms are designed to elicit that information. SSA may modify the report forms as long as the desired goal of ensuring that funds are used in the best interest of the beneficiary is met. Any changes to the report forms must be approved by the Office of Management and Budget.

However, any suggestions on how to revise the accounting forms within the meaning of the law would be welcome.

15. What can SSA do about nonreporting representative payees? Does SSA have any information on how it is enforcing this requirement?

SSA makes multiple attempts to obtain the required reports. These attempts may include additional mail requests, telephone contacts and re-directing the beneficiary's payment checks to a local field office to force the payee into the office so that annual reporting can be obtained. If all such efforts fail, SSA typically initiates a payee change. SSA does not track the number of payee changes made as a result of the failure to obtain a report after subsequent development efforts.

We understand that a significant number of the nonresponders are parents with custody of their own children and we believe that many do not respond because they believe we are being intrusive. For most of those cases, our experience has shown that it is *not* in the best interests of the child to remove the custodial parent as payee and replace them with an-other payee. We look forward to any analysis in the study with regard to whether nonresponders are more likely to misuse, and any suggestions you have on this subject.

POMS

16. It has been suggested that the POMS needs to have information and guidance on how to handle difficult situations, like homeless beneficia-ries, feuding family members, payees skimming off payments, payees not responding to attempts to contact them, and beneficiaries not wanting a payee when the staff thinks one is needed. Is SSA actively addressing these issues and if so, when will the new POMS be issued?

The POMS provides a wide variety of policy principles and instruc-tions, and attempts to cover most subject areas. The technicians need to use information contained in the POMS and use their own judgment and expertise to resolve the problems or issues presented to them in each par-ticular case. It would be impossible to cover by written instruction every conceivable situation the field will run into given the size and diversity of this population. However, the POMS does address most of the issues in your question as indicated below:

- "Payees skimming off payments" is misuse and is covered in GN 00604.
- "Payees not responding" is covered in GN 00605.
- "Beneficiaries not wanting a payee when SSA determines one is needed" is covered in GN 00504 and GN 00503.

All POMS sections are updated to provide clarifications and legislative changes as needed. Prior to publication, POMS are sent to all our regions for review and comment. Any areas that need additional clarification are brought to our attention at that time. In addition, whenever our regional contacts present ideas or suggestions to improve the POMS, we consider those ideas and issue changes as needed. If NAS has any particular suggestions for POMS enhancements, we welcome their input.

The Representative Payee System

17. **We were told that some offices are not using the RPS or that they do not follow all procedures in checking/entering data into the RPS. Do you have any information on this claim? What are the office directors' responsibilities in this regard?**

SSA policy states that, whenever possible, representative payee applications should be taken via the RPS as the RPS provides online safeguards. We believe that, generally, our field office components process representative payee applications cases via the RPS as instructed; however, we know that many technicians in the field have expressed frustration with the RPS saying that it is too rigid, cumbersome and error-prone. We also know that there are a significant number of cases missing from the RPS. Bypassing the RPS is of great concern to us because it prevents SSA from affording protections as designed. Because some rep payee applications cannot be taken in the RPS (undocumented alien parent payee without an SSN), we had to have processes in place so these cases could be processed outside of the RPS, a situation which also allows abuses to occur. Field office managers are ultimately responsible for the work performed by their employees.

While this creates a severe integrity issue for the representative payee program because the database is supposed to contain a record of all payees, implementing necessary systems changes is slow due to an antiquated database structure, resource constraints and other competing agency priorities.

18. **Does the RPS allow for a notation that the representative payee "may" be a misuser, or tends to be unresponsive to staff attempts to contact him or her? Is there any impediment to allowing this?**

If misuse is suspected, we must develop fully to make a determination whether the misuse actually occurred. Misuse is defined as "any case in which a representative payee converts the benefits for purposes other than the use and benefit of the beneficiary."

Once a misuse determination is made, RPS has a Representative Payee Misuse Information (RMIS) screen to record misuse information. This screen collects data about each misuse event and whether funds were reimbursed. We stress the importance of posting misuse information to RPS since this data will be displayed anytime a future application is filed by the payee or representative payee query is requested. We have included procedures in POMS for documenting misuse to the RPS.

RPS also has a Representative Payee Special Text (RPST) screen to house information that may be useful about the representative payee. This is a free-format screen and can be used for various events and could be used to document a payee who is unresponsive or who displays other indicators of potentially poor performance. It will be displayed anytime a future application is filed by the payee or a representative payee query is requested.

The RPS itself does not compile characteristics of payees and determine or note that someone is more likely to be a misuser. We look forward to your analysis of the common characteristics of a misuser so we can consider changes to the RPS or the accounting program.

19. How do staff suggestions for changes to the RPS filter upwards to headquarters? Our perception from talking to field staff is that ideas, such as "may be a misuser indicator," are stifled and no longer passed upwards.

Generally, we receive suggestions in one of two ways, either through our employee suggestion program or informally through our regional office staff. We want to make the representative payee process as easy and productive for the field offices as possible, so we welcome feedback on areas that need improvement. The agency encourages the employee suggestion program and we thoroughly consider all that are received. However, many suggestions are turned down because they do not provide sufficient benefits or savings to support the costs of implementing the change. In addition, due to limited systems resources, only a small number of systems changes are implemented each year within SSA.

20. Are there plans to upgrade the RPS to at least a Windows-based environment and to reduce the data entry duplication that field staff complain about and use as an excuse for not always using the system to its fullest potential?

Planning and analysis will start on upgrading the RPS to a more effective system later in FY 07 (from the current IDMS platform to a DB2 database). We do not know if funding will be available in order to implement this change but work has begun to identify the business process and some of the enhancements that should be incorporated.

Appendix C

Guardianship Questions to and Responses from the Social Security Administration

D uring the committee's deliberations several questions came up about guardianships, court-appointed guardianships, guardianship fees and representative payees. This appendix presents the committee's formal questions to the Social Security Administration (SSA) and the agency's written responses.

1. **When a court issues a guardianship and states that a guardian may bill for fees, may these fees be deducted from Social Security benefits?**

   Yes, the fees may be deducted from Social Security benefits.

2. **In a court-appointed guardian situation, is there any circumstance when fees are allowed or not allowed?**

   If the court approves a fee, SSA cannot deny its payment, and the use of Social Security benefits to pay court authorized guardianship fees is generally considered a proper use of the beneficiary's funds. However, if Social Security learns that a legal guardian is charging a fee that leaves the beneficiary with unmet needs, our policy provides that the case should be referred to the appropriate regional chief counsel (RCC) for review. If the RCC believes the charges are excessive, the circumstances should be brought to the court's attention. If representations to the court are unsuccessful, SSA will consider whether a change of payee is appropriate.

3. When a guardian calculates his/her fees based on a percentage of the total amount of income, may Social Security funds be included in the amount even if no Social Security funds are used to pay these fees?

Yes. Social Security does not instruct or guide the guardian payee in how to compute fees. As noted, SSA generally allows representative payees who are legal guardians to deduct court authorized guardianship fees and those fees may be deducted from Social Security benefits.

4. **May a guardian commingle Social Security benefits with other income or assets derived from sources that may include veteran's benefits, private pensions, or other routine deposits?**

SSA funds must be held in an account which is titled in such a manner as to be clear that the funds are owned by the beneficiary, not the payee. SSA has no restriction regarding holding Social Security funds along with other funds in the same account. Of course, the payee will have to account at least annually to SSA on the use of Social Security funds so, regardless of how the funds are held, a clear record must be kept regarding the use of Social Security funds.

5. **May a guardian bill for legal fees, tax preparation fees or other financial/legal processes where sources for these fees may be from Social Security funds?**

For court-approved fees, Social Security does not guide a guardian payee in how to compute a fee request that is submitted to the court.

6. **May State Guardianship Courts trump Social Security policy with regard to fees, when SSA policy states that only certain organizations may be paid a fee for service, and then it is capped on a monthly basis? State-appointed guardians are not fee-for-service organizations as defined by SSA.**

The guardianship fees allowed by State guardianship courts are not limited by Social Security policy. This is true because legal guardians are responsible for a wider range of duties and responsibilities than a Social Security representative payee who is not a legal guardian. However, a guardianship organization may also be approved by SSA to be a fee-for-service payee. If they provide evidence that they are performing different duties as a representative payee versus guardianship duties, they could collect a guardianship fee and a fee-for-service amount, but payment for the services may not overlap.

We welcome any recommendations you may have regarding the dual role of legal guardian and representative payee.

Appendix D

In-Depth Study of Misuse

The Social Security Administration (SSA) stores data about individual and organizational representative payees and their beneficiaries in the Representative Payee System (RPS). The system was created in response to the Omnibus Budget Reconciliation Act of 1990, which required SSA to develop and maintain a centralized file of representative payees (U.S. Social Security Administration, 2002).

For the payees and the beneficiaries, the RPS keeps data from the application to become a payee, certain demographic and socioeconomic characteristics, the benefit type, the history of payee-beneficiary relationships, and information about direct deposit. There is also an indicator of payee misuse. In March 2005, tallies generated from the RPS showed 11,464 representative payees identified as misusers—0.08 percent of the 14.3 million representative payees in the system at that time. If one assumes an equal annual distribution of misusers since 1992, when the RPS became operational, about 882 payees have been labeled misusers each year.

In order to learn if some payees are more likely to become misusers than others, we examined the RPS data and carried out an in-depth study on a small universe of misusers. The in-depth study was designed to focus on the documentation of misuse events, and it included a detailed review of beneficiary folders (which were retrieved from various storage locations). By understanding not only who is likely to be a misuser but also the circumstances involved in misuse, our goal was to determine if changes to policies and procedures are warranted to prevent future misuse in the program.

The first part of this study presents the results from our analysis of the RPS data. The second part presents the outcome of our in-depth review, focusing on how much money was misused, the length of the misuse, and the circumstances surrounding the misuse.

We stress that our analyses are confined by the limited nature of the existing variables in the RPS and by the very small number of identified misusers. As noted above, less than one-tenth of 1 percent of the payees in the RPS in 2005 were identified as misusers. Because of this very small number, the data in this study are descriptive: it was not possible for us to do any discriminant analyses with the available data.

ANALYSIS OF RPS DATA

In this section we first examine the characteristics of all the representative payees in the RPS by misuser status. We then discuss the characteristics of payees by misuser status and type of beneficiary. The last section discusses misuse status and relationship to beneficiary for currently active payees who are identified misusers.

Payees by Misuser Status

In order to learn if misusers have characteristics that are different from non-misusers, we created four dimensions for comparisons: (1) demographic characteristics, (2) stability in the community, (3) sources of income, and (4) indicators of criminal background.

Table D-1 shows a comparison of the payees on these dimensions. All representative payees with a history of misuse are compared with all other representative payees in the system. All comparisons except sex (p < .0370) and "other last name in RPS for the representative payee" (p < .2319) were statistically significant at the p < .0001 level.

Looking first at the demographic dimensions, misusers tend to be younger than non-misusers. In both groups, females are more likely to serve as representative payees than males, but there are slightly more males in the misuser group. The percentage of payees with different last names is about the same for the two groups (p < .2319).

With regard to indicators of stability in the community, misusers are less likely to use their residence address as their mailing address and more likely than non-misusers to have moved, both within the last year and for their time in the RPS. Misusers are more likely than non-misusers *not* to have provided a contact phone number.

For income, we selected six of the variables in the RPS for analysis:

TABLE D-1 Payee Characteristics by Misuser Status (in percentage)

Characteristic	Misusers[a]	Non-misusers[b]
Demographic Variables		
Under age 50	67.4	53.4
Male	28.7	27.8
Other last names in RPS	3.8	4.1
Stability in Community		
Same mailing and home address	84.3	88.0
Only one prior address in last year	59.3	77.6
Prior residence indicator	41.0	22.5
No telephone	4.9	2.3
Sources of Income		
Self-employment	3.4	1.9
Other than employment	38.8	24.2
OASDI or SSI	17.0	10.9
Public assistance	1.7	0.6
AFDC or TANF	6.1	1.6
Pension	1.4	2.4
Indicators of Criminal Background		
Ever convicted of a felony	6.2	2.4
Ever served time in prison	3.7	1.5

[a]N = 11,464.
[b]N = 14,380,000.

SOURCE: Data generated from a review of SSA administrative data of identified misusers conducted for the National Academies Committee on Social Security Representative Payees (2006).

self-employment, income from sources other than employment, receipt of Social Security benefits, income from public assistance, income from federal welfare programs,[1] and income from pensions.

Table D-1 presents the data on several demographic variables and on income. Self-employment is more likely for misusers than non-misusers, and misusers are also more likely to have income from sources other than employment. Those other sources include Social Security benefits—Old Age Survivors and Disability Insurance (OASDI), Supplemental Security Income (SSI), or both. More misusers than non-misusers say public assistance is a source of their income. Finally, the percentage who reported welfare benefits as a source of income is higher among misusers than non-misusers. The percentage with a pension as a source of income is lower for misusers

---

[1]Until 1997, the primary federal welfare program was Aid to Families with Dependent Children (AFDC); since that date it has been Temporary Assistance for Needy Families (TANF).

than for non-misusers, perhaps reflecting the younger age distribution of the misusers.

Finally, on the fourth dimension, Table D-1 shows that misusers are more likely than non-misusers to have been convicted of a felony or to have served time in prison. For misusers, the percentage who have served time in prison is higher than the national average: 3.7 percent compared with 2.7 (see www.ojp.usdoj.gov/bjs/crimoff.htm).

Payees by Misuser Status and Type of Beneficiary

We examined the data to learn if the misuser payees are serving different types of beneficiaries than the non-misusers. Specifically, we wanted to know if family membership is related to misuse. For this analysis, we categorized beneficiaries in three groups: family members, nonfamily members, and a mix of family members and nonfamily members.

Table D-2 presents the data on these variables. Almost all the representative payees serve family members only. In general, there is more misuse for representative payees who serve both family members and nonfamily members (i.e., they have a mix of beneficiaries). The misusers are younger than the non-misusers across the three beneficiary types. Misusers are less stable in their communities than non-misusers as measured by the percentage with only one prior address in the last year or the percentage with a prior residence ever. The representative payees with a mix of beneficiary types (family and nonfamily members) tend to be the most mobile. Misusers with a mix of family and nonfamily members are also more likely to have income from sources other than employment, including OASDI and SSI benefits, welfare (AFDC or TANF), public assistance, and pensions. Misusers with family members as beneficiaries are less likely to have been convicted of a felony or served time in prison than misusers with nonfamily members or a mix of beneficiaries.

Active Payees Identified as Misusers

The analysis of the RPS records revealed that some payees who have been identified as misusers have not been terminated as payees: they continue to serve either for the beneficiary whose funds they misused or for others. Of the approximately 5.3 million active representative payees in March 2005, 2,359 or 0.04 percent carry the label "previous misuser." They serve 3,620 beneficiaries.

Table D-3 shows the relationship of active payees to their beneficiaries by misuser status. Almost 55 percent of the non-misusers are mothers, but among the misusers the percentage is 61.3 percent (0.06 percent of mothers are misusers). About 3 percent of all non-misuser payees are grandparents,

TABLE D-2 Payee Characteristics by Misuser Status and Type of Beneficiary (in percentage)

Characteristics	Family Only		Nonfamily		Mix	
	Misuser[a]	Non-misuser[b]	Misuser[c]	Non-misuser[d]	Misuser[e]	Non-misuser[f]
Demographic Variables						
Under age 50	66.5	50.9	58.5	44.5	59.6	50.8
Male	27.2	26.9	48.3	43.9	20.0	21.2
Other last names in RPS	4.1	4.1	3.8	4.1	2.0	2.8
Stability in Community						
Same mailing and home address	85.1	89.0	82.2	83.0	81.0	85.2
Only one prior address in last year	60.2	77.9	65.2	78.2	47.8	59.9
Prior residence indicator	40.0	22.2	35.2	21.9	52.6	40.3
No telephone	4.6	2.1	6.2	4.0	5.2	3.5
Sources of Income						
Self-employment	1.9	1.4	12.2	8.8	5.4	4.6
Other than employment	31.4	22.0	56.9	50.9	70.1	63.1
OASDI or SSI	13.9	10.0	15.8	18.5	38.6	33.5
Public assistance	1.4	0.4	3.5	1.9	2.4	2.3
AFDC or TANF	5.0	1.3	6.0	3.4	12.6	8.8
Pension	1.1	2.3	2.2	4.2	2.4	3.3
Indicators of Criminal Background						
Ever convicted of a felony	6.0	2.3	7.3	4.2	7.1	4.5
Ever served time in prison	3.5	1.4	4.6	2.9	3.9	2.7

[a]N = 8,837.
[b]N = 13,300,000.
[c]N = 1,255.
[d]N = 725,080.
[e]N = 1,354.
[f]N = 355,980.

SOURCE: Data generated from a review of SSA administrative data of identified misusers conducted for the National Academies Committee on Social Security Representative Payees (2006).

TABLE D-3 Active Representative Payee by Relationship to Beneficiary
and Misuser Status (in percentage)

Relationship	Percent Misusers	Percent of All	
		Misusers	Non-misusers
Mother	0.06	61.3	54.5
Other relative	0.05	9.4	9.9
Self	0.07	9.4	7.3
Grandparent	0.09	6.1	3.4
Other	0.07	5.0	3.4
Father	0.02	4.4	13.2
Child	0.03	2.2	4.2
Spouse	0.03	2.2	3.6
Stepfather	0.00	0.0	0.3
Stepmother	0.00	0.0	0.2

SOURCE: Data generated from a review of SSA administrative data of identified misusers conducted for the National Academies Committee on Social Security Representative Payees (2006).

but among the active payees identified as misusers the percentage of grand-parents is 6.1 (0.09 percent of grandparents are misusers). In comparison, about 13 percent of non-misusers are fathers, but their percentage of the active payees identified as misusers is only 4.4 (0.02 percent of payee fathers are misusers). Thus the percent of mothers and grandparents who are mis-users is proportionally higher than the percent that are non-misusers.

It is important to note that we could not discern from the available RPS data whether the active misuser payees: (1) continue to serve for the beneficiary whose funds they had misused, (2) were terminated as the payee for the beneficiary whose funds were misused, but continue to serve for other beneficiaries, or (3) were terminated as the payee for the beneficiary whose funds were misused and then subsequently selected for the same or other beneficiaries.

We randomly selected 10 active misuser payees (and their beneficiaries) in the RPS for further analysis, hoping that a few anecdotes would pro-vide some insight into this universe of payees, but we did not see anything unique about this group. Among the 10 cases, we found mothers who are continuing to serve for their child or children whose funds they misused, and mothers serving different children. One mother—who has two sets of children far apart in ages living in different states—was terminated as payee for her older children, but subsequently appointed to serve for her younger children. One payee who misused the benefits for her ex-spouse is now the payee for her child. In two other cases a brother continues to serve, though he misused another brother's benefits and an unrelated payee is still serving for other unrelated beneficiaries. In one case the RPS notes suggest that the misuse is only alleged and that the designation of misuse might be inaccurate: the payee is an adult daughter who continues to serve for the same beneficiary.

Summary

Our analysis of the RPS data provides some information about the type of payee who is identified as a misuser in the system. Compared with non-misusers, the misusers tend to

- be younger;
- be less likely to have the same mailing and residence addresses;
- move more frequently;
- be self-employed;
- have income from sources other than employment;
- be recipients of public assistance;
- be OASDI or SSI beneficiaries themselves; and
- have a criminal background.

When relationship to beneficiary is taken into account, we find the following:

- In general, payees tend to serve family members only, but misusers are disproportionately found among payees who serve a mix of family and nonfamily members.
- In general, payees tend to be female, whether misusers or not, but misusers who serve nonfamily members tend to be disproportionately male.
- Misusers tend to have income from sources other than employment; this tendency is especially strong for misusers who serve for a mix of family and nonfamily members.
- As noted above, misusers are more likely to have criminal backgrounds than non-misusers, and this tendency is especially strong for payees who serve nonfamily members or a mix of family and nonfamily members.

The analysis showed that more than 2,000 active payees have a prior history of misuse. Mothers and grandparents make up an even higher percentage of these payees than of all payees, while fathers and spouses make up a lower percentage. Payees with a history of misuse have been allowed to continue to serve as payees for either the beneficiary whose funds they misused or for other beneficiaries for whom they have no record of misuse.

IN-DEPTH ANALYSIS

Some information about the circumstances surrounding the misuse can be gleaned from notes in the RPS, but the picture is incomplete. Therefore, the committee carried out an in-depth analysis in order to learn more about misuser representative payees. We compiled data that would help us describe the misuse, the factors leading up to the misuse determination, the amount of money involved, the characteristics of the payees and the beneficiaries at the time of the misuse, including the beneficiaries' living situations and relationships to their payees.

Process

We began the in-depth analysis with a pilot review of 10 randomly selected beneficiary folders of representative payees identified in the RPS as misusers and serving beneficiaries with disabilities. The folders for these beneficiaries were stored at the SSA Processing Service Center in Baltimore and therefore could be retrieved fairly quickly. This initial review was designed to determine what information could be retrieved from the beneficiary folders and what information we would need to retrieve from other administrative records, including the RPS, to supplement the contents of the folders.

In addition to RPS and the 10 pilot folders, we used the following SSA data files for this analysis:

ROAR   Recovery of Overpayments, Accounting and Reporting File
MBR   Master Beneficiary File
SSID   Supplemental Security Income Display
SEQY   Summary of Earnings Query
NUMI   Numident File
PUPS   Prisoner Update Processing System

From the information available from all the sources—and after appropriate input from staff of the Representative Payee Program—the committee developed a 60-item recording sheet that also included ample space for the case reviewers to provide information about the misuse event.

SSA staff members assisted with the request of folders. They also set up a system to track the folders through the review, to store the folders temporarily at SSA HQ and to return the folders to storage after the review. The SSA Representative Payee Program staff performed the actual case review under the guidance of NAS staff.

Two training sessions were held for the case reviewers, which led to a few adjustments to the case review process and further refinements to the recording sheets. For many of the payees and beneficiaries, the case reviewers had to read various notes in several folders in order to glean the information presented in this study.

To meet our goal of reviewing 300 folders, we drew a random sample of 500 non-fee-for-service individual representative payees from a list we had generated from the RPS of previously identified misusers. Because the RPS dates to 1992, some of the beneficiary folders were old. Of the 500 beneficiary folders initially requested, 206 had been destroyed or were not in central storage (stored instead in a field office or a payment center), and another 26 were determined by the Representative Payee Program staff to have been erroneously labeled as misusers in the RPS. Thus, we had to draw a supplementary sample to ensure we had about 300 folders to review.

Ultimately, we obtained and reviewed 291 folders, which represents 2.54 percent of the RPS misuser payee universe (11,464).

Characteristics of Misusers and Their Beneficiaries

In order to ensure that the payees selected for the in-depth study were representative of the larger universe of misuser payees (described above), we examined and confirmed the distribution of their characteristics on the indicators shown in Tables D-1 to D-3. In the process, we also looked for additional characteristics, such as education, that might be useful in understanding misusers.

Demographic Characteristics: Payees and Their Beneficiaries

Table D-4 shows the demographic characteristics of the payees (at the time of the misuse). More than 65 percent of the misuser payees are female, and more than 50 percent of the payees are between the ages of 30 and 50. Most of the payees are the parents of the beneficiaries. Table D-5 shows the demographic characteristics of the beneficiaries of the payee misusers. The beneficiaries tend to be male and minors. Consistent with the findings for the payees, most beneficiaries are the sons or daughters of the payees.

Custody and Guardianship Arrangements

We looked at the custody arrangements of the payees and their beneficiaries at two points in time: application and occasion of misuse. At the time of the application to become the representative payee, of the 291 cases, 42 percent of the payees who became misusers (123) had physical custody of the beneficiary; 22 percent did not (65); the information was not known or available for 35 percent (103).

At the time of the misuse, 16 percent (48) of payees had physical cus-

TABLE D-4 Payee Demographic Characteristics at Time of Misuse

Characteristic	Number	Percent
Gender		
Male	84	28.9
Female	190	65.3
Not available	17	5.8
Age		
19	1	0.3
20-29	44	15.1
30-39	108	37.1
40-49	64	22.0
50-59	32	11.0
60-69	8	2.8
70 and older	5	1.7
Not available	29	10.0
Age Relative to the Beneficiary's Age		
Younger	47	16.2
Older	207	71.1
Not available	37	12.7
Relationship to Beneficiary		
Spouse	9	3.1
Parent	146	50.2
Son or daughter	8	2.8
Grandparent	2	0.7
Grandchild	1	0.3
Other relative	36	12.4
Other	24	8.5
Not available	65	22.3

SOURCE: Data generated from a review of SSA administrative data of identified misusers conducted for the National Academies Committee on Social Security Representative Payees (2006).

tody of their beneficiaries; 46 percent did not (133); and the information was not known or available for 38 percent (110) of the payees.

Of those who indicated on the application form that they had physical custody of the beneficiary, just under 33 percent had physical custody at the time of the misuse (40 of 123); and 53 percent no longer had physical custody (65); and for 15 percent the information was unavailable or unknown. A very small fraction, 6 percent (4 of 65), did not have custody at the time of the application, but they did at the time of misuse.

Thus, for the payees for whom information is available, it appears that many payees who misuse funds do not have the beneficiary in their custody at the time of misuse although they did at the time of the application.

With regard to guardianship, information was not available for the time at which the payees applied to become or became the payees. At the

TABLE D-5 Beneficiary Demographic Characteristics at Time of Misuse

Characteristic	Number	Percent
Gender		
Male	163	56.0
Female	107	35.1
Not available	8	8.9
Age		
Under 18	160	55.0
18-64	107	36.8
65 and older	8	2.8
Not available	16	5.5
Relationship to Payee		
Spouse	8	2.8
Parent	16	5.5
Son/daughter	136	46.7
Grandparent	1	0.3
Grandchild	4	1.4
Other relative	35	12.0
Other	24	8.3
Not available	67	23.0

SOURCE: Data generated from a review of SSA administrative data of identified misusers conducted for the National Academies Committee on Social Security Representative Payees (2006).

time of the misuse, only 4 percent of the payees were court-appointed legal guardians of their payees; 73 percent were not, and for 23 percent the information is not available.

Socioeconomic Characteristics: Payees and Their Beneficiaries

Table D-6 shows some education and income characteristics of payees at the time of the misuse. Information on payees' level of education is not formally included in any of the SSA administrative data: the case reviewers were able to obtain the information by reading through the materials in the review folders looking for any notes with reference to this item. From the information they were able to pull together, it appears that at the time of the misuse only a small portion of the payees had a high school or higher level of education; however, data were not available for 83 percent of the payees.

For sources of income and income amount, data were available for almost 75 percent of payees. For those payees, the two major sources of income—slightly less than one-fourth each—are employment and OASDI and more than two-thirds had annual incomes of $20,000 or less.

Slightly more than one-third of the payees (101) were themselves beneficiaries at the time of the misuse. Of them, close to 60 percent (59) re-

TABLE D-6 Payee Education, Sources of Income, and Income at Time of Misuse

Characteristic	Number	Percent
Education		
No formal schooling	0	0.0
Less than high school	39	13.4
High school	8	2.8
Post high school	2	0.7
Not available	242	83.2
Source of Income		
Employed	67	23.0
Self-employed	4	1.4
OASDI	68	23.4
Pension	1	0.3
SSI	17	5.8
AFDC or TANF	11	3.8
Employed and OASDI	6	2.1
Employed and SSI	3	1.0
Employed and other	1	0.3
Self-employed and AFDC or TANF	4	1.4
OASDI and AFDC or TANF	11	3.8
OASDI and SSI	9	3.1
OASDI and worker's compensation	1	0.3
Employed and OASDI and other	1	0.3
Employed and AFDC or TANF and other	1	0.3
Other	10	3.4
Unknown	76	26.1
Annual Income from All Sources		
Less than $5,000	68	23.4
5,000-10,000	69	23.7
10,000-20,000	58	19.9
20,000-30,000	11	3.8
30,000-40,000	5	1.7
40,000-50,000	1	0.3
More than $50,000	2	0.7
Unknown	77	26.5

SOURCE: Data generated from a review of SSA administrative data of identified misusers conducted for the National Academies Committee on Social Security Representative Payees (2006).

ceived benefits because of a disability; almost 20 percent (19) received SSI; and another 20 percent received OASDI benefits.

Table D-7 shows information on beneficiaries' education and type of benefit at the time of misuse. Since most of the misuse involves minors, it is not surprising that many have not finished high school. Most of the beneficiaries were receiving OASDI disability or SSI benefits.

TABLE D-7 Beneficiary Education and Benefits at Time of Misuse

Characteristic	Number	Percent
Education		
No formal schooling	10	3.4
Still in school	64	22.0
Less than high school	88	30.2
High school	16	5.5
Post-high school	10	3.4
Not available	73	35.4
Type of Benefit Received		
OASDI: Retirement	6	2.1
OASDI: Survivors	40	13.8
OASDI: Disability	91	31.3
SSI	101	34.7
Concurrent	23	7.9
Not available	30	10.3

SOURCE: Data generated from a review of SSA administrative data of identified misusers conducted for the National Academies Committee on Social Security Representative Payees (2006).

Community Stability and Language: Payees and Their Beneficiaries

Table D-8 shows some characteristics related to stability in the community of the payees and their beneficiaries. About 12 percent of payees had more than one address in a year; of their beneficiaries, 11 percent had moved frequently. More than 80 percent of payees provided a phone number at the time of the application (although there is no information to verify if this was the correct number). Close to 72 percent of payees gave the same residence and mailing address.

We also looked at language spoken and used for communication between payees and their beneficiaries. Of the 83 percent (242) for whom this information was available, 82 percent (238) spoke the same language.

Criminal and Drug or Alcohol Background: Payees and Their Beneficiaries

About 5.5 percent (16 of 291) of the payees had a criminal record. For comparison, in the larger universe of misusers, 6 percent of the payees had been convicted of a felony, and 3.7 percent had served time in prison (see Table D-1).

Table D-9 presents information on drug or alcohol abuse by payees or their beneficiaries. For both payees and beneficiaries, information in the case folders indicated only small fractions were abusing drugs or alcohol.

TABLE D-8 Payee and Beneficiary Stability in the Community

Characteristic	Number	Percent
Payee: More Than One Address in a Year		
Yes	34	11.7
No	138	47.4
Not available	119	40.9
Payee: Provided a Telephone Number		
Yes	238	81.8
No	27	9.3
Not available	26	8.9
Payee: Same Residence and Mailing Address		
Yes	209	71.8
No	18	6.2
Not available	64	22.0
Beneficiary: Moves Frequently		
Yes	32	11.0
No	182	62.5
Not available	77	26.5

SOURCE: Data generated from a review of SSA administrative data of identified misusers conducted for the National Academies Committee on Social Security Representative Payees (2006).

TABLE D-9 Payee and Beneficiary Reported Drug or Alcohol Abuse

Drug or Alcohol Abuse	Number	Percent
Payees		
Known to abuse	20	6.9
Reviewers found no allegations	150	51.6
Not available	121	41.6
Beneficiaries		
Known to abuse	13	4.5
Reviewers found no allegations	209	71.8
Not available	69	23.7

SOURCE: Data generated from a review of SSA administrative data of identified misusers conducted for the National Academies Committee on Social Security Representative Payees (2006).

Mental Health and Competence: Beneficiaries

Of the beneficiaries, 25 percent (73) were alleged to be suffering from a mental illness; 53 percent (155) were not; and for 22 percent (63) the information was not available. Only a very small fraction of the beneficiaries, less than 2 percent (4) had been judged to be legally incompetent.

Other Factors Related to Payeeship

Table D-10 presents data on a number of other factors retrievable from the RPS, the review folders, or other SSA administrative data files that can be classified under the heading payeeship. The case review revealed that one payee himself had a payee. Almost 25 percent of the payees were serving for several beneficiaries at the time of misuse. About 17 percent were on record as having failed to respond to the annual accounting process.

In spite of being labeled a misuser in the RPS, about 9 percent of the payees were still shown as active in the system. This percentage is somewhat

TABLE D-10 Factors Related to the Payeeship

Payee	Number	Percent
Payee Is a Payee		
Yes	1	0.3
No	272	93.5
Not available	18	6.2
Responds to Annual Accounting		
Yes	25	8.6
No	49	16.8
Not available	217	74.6
Serving for Others at Time of Misuse		
Yes	70	24.1
No	170	58.4
Not available	51	17.5
Still Serving as Payee		
Yes	27	9.3
No	234	80.4
Not available	30	10.3
Total Number of Beneficiaries Served		
Only one	102	35.1
2-15	159	54.6
15-50	2	0.7
50 or more	12	4.1
Not available	16	5.5
Rapid Increase in Number of Beneficiaries Served		
Yes	2	0.7
No	246	84.5
Not available	43	14.8
Number of Times Terminated		
Once	129	44.3
Twice	47	16.2
More than twice	64	22.0
Never	11	3.8
Not available	40	13.8

SOURCE: Data generated from a review of SSA administrative data of identified misusers conducted for the National Academies Committee on Social Security Representative Payees (2006).

lower than the percentage in the larger universe: about 20 percent of the 11,464 identified misusers in the RPS were shown to be active.

In the RPS, a notation of "more suitable payee" is frequently used as the explanation when there is a payee change. The case review showed that about 44 percent of the misuser payees had been terminated once, 16 percent twice, and 22 percent more than twice because "a more suitable payee" was found for either the beneficiary associated with the misuse or for other beneficiaries served by this payee.

In summary, it appears that the cases selected for the in-depth study are similar to the larger universe of identified misusers in the RPS.

Characteristics of the Misuse

Amount Misused and Length of Misuse

Table D-11 shows the amounts misused by the payees. Data were available for 273 of the 291 payees in the in-depth study. As shown in the table, the total amount misused was close to $1.2 million. The average misuse amount was less than $4,500; the smallest amount misused was $41 for 1 month, and the highest amount misused was $45,000 over 9 years and 4 months. The longest period of misuse was more than 10 years and involved a total of $33,000.

TABLE D-11 Misuse Amounts and Time

Category	Unit
Amount	
Total	$1,191,859
Average	$4,414
Median	$2,442
Lowest	$41
Highest	$45,000
Duration in Months	
Average time	11 months
Median	4 months
Shortest	1 month
Longest	129 months

SOURCE: Data generated from a review of SSA administrative data of identified misusers conducted for the National Academies Committee on Social Security Representative Payees (2006).

TABLE D-12 Distribution of Duration of
Misuse (in percentage)

Length of Time	Percent
Months	
1	35.0
2	6.9
3	4.4
4	5.5
5	3.3
6	3.7
7	3.7
8	4.0
9	2.2
10	2.6
11	2.9
12	3.7
Total	77.7
Years (in months)	
0-12	77.7
12-24	10.3
24-36	4.4
36-48	3.7
48-60	1.5
60-72	0.7
72-84	0.7
84-96	0.4
96-108	0.0
108-120	0.4
120-132	0.4
Total	100.0

SOURCE: Data generated from a review of SSA administrative data of identified misusers conducted for the National Academies Committee on Social Security Representative Payees (2006).

Table D-12 shows the percent distribution of the misuse duration in two ways: for the first 12 months and by number of years (in months). Most misuse involves just 1 month of benefits (35 percent), and more than 77 percent lasts no more than 12 months.

Direct Deposit

SSA strongly encourages all beneficiaries to receive their monthly benefits by direct deposit. As of October 2005, 83.1 percent of *all* OASDI and

56.2 percent of *all* SSI beneficiaries received their benefits by direct deposit (see http://www.ssa.gov/deposit/GIS/data/Reports/DDTREND2.htm [June 2007]). There is no official statistic for beneficiaries with representative payee. From the case review, it appears that 18 percent of the beneficiaries had direct deposit at the time of the misuse.

SSA Negligence and Recovery of Funds

SSA rarely finds itself negligent, i.e., to blame for creating the misuse situation. The in-depth study uncovered only three examples where SSA acknowledged that it should not have provided the check to the payee: (1) SSA incorrectly sent a check to the father, though he had *not* filed to be the payee (the mother had filed); (2) SSA failed to stop a check from going to the wrong payee; and (3) SSA gave a special immediate payment check to a mother, even though the agency knew she did not have the beneficiary in her custody.

If misuse has been established or an overpayment has been made and SSA does not find itself negligent, the payee can be made to pay back the misused or overpaid amount. Such payments can be withheld from the payee's own current or future benefit checks. The process is illustrated in the remarks made by one of the case reviewers:

> The three kids were not in their mother's custody off and on over a four-year period. The computer records do not reflect the true overpayment or misuse amount. According to the Office of the Inspector General summary report of investigation the mother was overpaid $17,562.60 due to not having custody of the kids and misusing the funds. She subsequently became entitled to disabled widower's benefits and her retro-benefits of $14,665.60 were used to reduce the misuse amount to $2,897.00. This is being withheld from her benefits at $40.80 per month.

Table D-13 shows recovery information by whether or not the payee disputed the misuse allegation for the 125 cases for which the information is available. Misused benefit funds were recovered for less than 1 percent if the payee disputed the misuse determination and for 10.3 percent if there was no dispute. For those 125 cases, information on actual recovery was available for 103 of them: funds were recovered for less than one-third of them (32).

Explanations of the Misuse

The case reviewers looked for documentation that would shed light on the circumstances surrounding the misuse. They searched for narratives about the misuse events and how SSA discovered the misuse. Table D-14

TABLE D-13 Recovery of Misused Money by Payee's Decision to Dispute

Payee Dispute	Misuse Recovered	Number	Percent
Yes	Yes	2	0.7
	No	6	2.1
	Don't know	2	0.7
	Total	10	3.5
No			
	Yes	30	10.3
	No	65	22.3
	Don't know	20	6.9
	Total	115	39.5
Not Known		166	57.0
		291	100.0

SOURCE: Data generated from a review of SSA administrative data of identified misusers conducted for the National Academies Committee on Social Security Representative Payees (2006).

presents the distribution of misuse events by categories of explanation. We note above and in the body of the report that misuse is not well documented in the RPS, and it is also not well documented in the beneficiary folders or other administrative data. As shown in this table, the case reviewers were able to piece together explanations or find documentation for only 59 percent of the cases (172 of 291).

For about 16 percent of these cases, the explanations of misuse are straightforward, as illustrated by the quotes found in the beneficiary folders or provided by the SSA staff who reviewed the cases:

- The payee takes the money, and does not spend it on the beneficiary.
- Did not use funds for beneficiary.
- Beneficiary moved and payee kept the money. Payee was the deceased mother's boyfriend.
- Beneficiary reported payee not using money for her care. Her rent was allegedly not paid.
- Beneficiary claims that daughter took off with money (one month's benefit) and left town.
- Payee left home and used benefit check for self.
- Beneficiary applied to be her own payee and reported never receiving any of the money payee received.
- Payee used benefits for self rather than beneficiary. Payee would claim nonreceipt of beneficiary's check and then cash reissued check as well as original check.

TABLE D-14 Explanations of Misuse Events by Categories of Explanations

Explanation	Number	Percent	Number	Percent
Yes	172	59.1		
Miscellaneous			28	16.3
Physical custody				
Parent			77	44.8
Other payee			<u>13</u>	<u>7.6</u>
Total			90	52.4
Payee issues				
Payee could not be located by SSA			2	1.2
Payee was not the payee of record			4	2.3
Payee was in prison while listed as payee			4	2.3
Payee could not account for spending			<u>6</u>	<u>3.5</u>
Total			16	9.3
Misuse/inappropriate spending				
Payee misused dedicated account			9	5.2
Benefits not due beneficiary				
Beneficiary was deceased			1	0.6
Beneficiary was in prison/institutionalized			12	7.0
Beneficiary was not due the money - other			<u>9</u>	<u>5.2</u>
Total			22	12.8
Misuse was alleged				
Misuse was never substantiated			7	4.1
No	<u>119</u>	<u>40.9</u>		
Total	291	100.0		

SOURCE: Data generated from a review of SSA administrative data of identified misusers

- Payee reports not receiving check and cashes both checks—double check abuser.

By far the largest category of misuse determination involves SSA's unknowingly sending the benefit check to a person, who does not have the beneficiary in his or her care (52 percent).

Within the physical custody category, narratives and anecdotes suggest that much of the known misuse involves custodial disputes between parents of minor children. A typical scenario in this category is for the payments for minor beneficiaries to go to the father when the mother has physical custody of them. If other relatives are involved, the most likely scenario is for the payments to go to a mother or a father when they should be going to a custodial grandparent. The following direct quotes illustrate the custody issues:

- Father received a large retro-check for son in the amount of $7,197. On 1/99 payee changed to mother effective 2/99. Father never sent money to new payee (mother). Parents had separated on 12/93 and mother has had custody the entire time.
- Father received benefits for children on deceased mother's records. Children lived with grandmother, who was later granted legal and physical custody.
- Father used the money on himself. He did not have child in his care.
- Mother lost custody and did not use money for kids.
- Child actually lived with father while mother was payee—mother did not report custody change.
- Client's mother filed to be payee in 2001 stating she had physical custody of her son. In 2005 client's father found out that the mother had been collecting benefits for the child. Father was able to prove that he has had full legal and physical custody of his son since 1999. Mother misused monies received from 2001 to 2004.
- Misuser payee alleged custody; beneficiary's sister initiated investigation; misuser did not have custody and did not use funds for beneficiary.
- Beneficiary went into foster care and payee kept the money for self.
- Third party reported that children have not lived with mother for three years. Children are in foster care.
- Payee did not have beneficiary in her custody. Failed to report beneficiary moved in with father.

Sometimes the payees themselves feel that they are entitled to the beneficiary's money even though they do not have the beneficiary in physical custody, as in this example:

Beneficiary started working, got married, was no longer in payee's care and was not due the money. Payee was a mother with four children. She looked at the monthly benefit as a family allotment, not as separate money for each child. She was found to misuse close to $9,000 over 26 months. She "forgot" to tell SSA, that "Donald," age 16, moved out, got married and started working. Payee said: "Donald removed himself from the family. He was welcome to return to our house at any time, and I expected that he would do that. However, my wishes were to be in vain, and the days stretched into weeks, weeks into months and months into years. Furthermore, to have divided up the benefits from SSA and given to Donald the amount you claim was due to him, would have left the other children in need."

Smaller categories of explanation for misuse include the inability of SSA to locate the payee (1 percent), payee not being the payee on record (2 percent), and as illustrated by the narrative below, payee being in prison, yet still serving as payee (2 percent).

The payee was incarcerated and had her mother cash the beneficiary's checks. Some of the money was used to pay her bail and court expenses.

According to case reviewers, the RPS misuse indicator was sometimes assigned because the payee failed or refused to participate in the annual accounting process (3 percent). In other words, it appears that SSA sometimes assumed misuse because the accounting did not take place, as illustrated in this statement in the records:

- Payee did not respond to concerns of field office about payee accounting. Accounting report raised questions.
- Payee would not give information regarding beneficiary's conserved funds.
- Payee has not responded to multiple requests for an accounting of the benefits he received.
- Payee refused to give nursing home money and refused to account.

The misuse indicator may also reflect that the payee spent the money in a dedicated account on purchases that do not benefit the beneficiary or for nonauthorized expenditures. About 5 percent of the cases are classified this way. In these cases, judging from the SSA staff reviewer notes, there is some ambiguity on their part in determining what constitutes misuse and what is inappropriate or unauthorized spending. The SSA staff made the following type of observations:

- Misapplied monies from dedicated account, not misuse.
- This case involves misapplication of dedicated account funds which differ from misuse. However, the field staff posted this as misuse on RPS in error.
- RP [payee] mismanaged dedicated account funds and also used funds in an unauthorized manner (to repay a personal loan).
- A note in the folder indicates that $196 was misapplied and $387 was misused by payee. Retro-money for beneficiary was supposed

to be in a dedicated account to be used for specific things. Payee used it for things other than what qualifies, and it was considered misuse although payee was not charged.

If the beneficiary is in prison, institutionalized, or deceased, SSA is supposed to be notified and the monthly benefits should be suspended or cease. About 7 percent of the misuse determinations were in this category. Such infractions are sometimes noted to be overpayments to the payees. Judging from the case reviewers' comments, there appears to be some ambivalence about calling this misuse (see Chapter 5), as illustrated in these remarks:

- The payee (parent) cashed two checks after death of beneficiary (daughter).
- Payee failed to report that beneficiary was in jail. Payee cashed checks despite beneficiary being in an institution.
- Beneficiary was in jail. Payee withdrew all monies from account and would not account for it.
- Payee reported fraudulent change of address for beneficiary who was in jail.
- Overpayment. Payee did not report that beneficiary is incarcerated.
- Overpayment. Beneficiary is in prison and did not report.
- Overpayment due to prison.
- Not a case of misuse. Strictly an overpayment case. Why? From 6/99-8/99 beneficiary was entitled to no money because of being in an institution.

There are other reasons that the beneficiary should not have been receiving a monthly benefit. Again, the case reviewers found that many of these situations should have been characterized as overpayments, not misuse:

- This is an overpayment because mom won lottery and on another occasion got $28,000 in insurance money. Therefore, beneficiary was not due checks.
- Beneficiary was ineligible to receive the Title 16 benefits. He was working, thus overpaid.
- This is an overpayment, because beneficiary sold a house for $61,000 and did not qualify for SSI benefits.
- The SSI redetermination was denied. Therefore, beneficiary is overpaid for those months. She is additionally overpaid because for 7 months during the above period, duplicate checks were issued and cashed.

- This is not a misuse case. It is an overpayment. The payee failed to report beneficiary's disability had improved.
- Apparently, payee was overpaid because she received the benefits (SSI) on behalf of the beneficiary and later beneficiary was found ineligible. Payee requested reconsideration and continuance of SSI. Administrative Law Judge found benefits should terminate.

The case reviewers found notes to suggest that about 4 percent of the misuse cases were actually never substantiated and in their opinion would have been more properly classified as *alleged misuse*:

- Misuse was originally alleged by recipient against his payee/mother. Field office did preliminary development and determined the allegation was without merit. Therefore, misuse proceedings were abandoned.
- Payee alleges beneficiary took money. There is no evidence of that in the folder.
- No indication what misuse was about. The records indicated that number holder was in nursing home and that payee was still receiving checks, but alleged that checks were not coming.
- No misuse evidence in file or on RPS other than payee abused beneficiary. No determination available. Misuse reported by adoptive father of beneficiary.
- The misuse determination found there was no misuse. Payee's spouse, beneficiary's mother was complainant. She has a history of alleging misuse. Payee presented receipts and listed expenses.
- Beneficiary filed to be her own payee at age 18 claiming former payee misused her SSI money—no other evidence regarding misuse is available.

At times, the case reviewers found that the field office made a misuse notation in the RPS even though the field office investigation suggested that no misuse took place, as exemplified in this case:

- On 4/16/04, a third party contacted SSA complaining that the RP [payee] was borrowing money and not returning it to her. The claims representative did an alpha search on X and found the payee, Y and investigated this payee for misuse. On 5/10/04 the payee was contacted and asked to come to the office with all receipts and bank statements for X's SSI payments. The payee brought every receipt and bank statement. The payee admitted that from time to time he would borrow money from beneficiary and repay it the same day or very next. RP kept excellent records of beneficiary's SSI payments. Field office contacted beneficiary for a statement concerning the payee's suitability and whether he wished the payee to continue. The statement indicated the RP borrows money, but pays it back and beneficiary had no problem with this and had given RP the approval to borrow funds. The FO found no misuse and continued RP with the beneficiary's approval.

Sometimes, the case reviewers questioned the accuracy of the misuse notation in the RPS:

- This was originally coded as misuse over 19 months in the amount of $4,767. As of 8/30/05, the RPS now has the code "not a misuser." There is no information about misuse in the folder.
- Misuse indicator posted to the wrong payee. The records for this payee do not indicate misuse.
- It cannot be misuse, if the amount is zero.
- Beneficiary is his own payee. One cannot be found to misuse own benefits.
- RPS remarks show: The mother was her payee in the past. T2 claim was locked, so they showed misuse to clear it. There was no misuse.

For a total of 41 percent of the cases, no explanation could be found. That is, if the SSA staff who reviewed the folders could not find any explanation of the misuse or could not deduct a reason from the review, they would make a note of "unknown" or "no data in the files." Their comments are illustrative of these cases:

- Beneficiary's RPS screen shows payee changed for more suitable payee. The beneficiary applied to become his own payee. Nothing in the file makes reference to the misuse issue.
- There is no information pertaining to misuse in the claims folder. Perhaps misuse can be assumed by looking at the RPS screens, which show a payee change in the period the misuse occurred.
- Not sure if misuse ever happened. There is nothing on queries to verify it, and no misuse data in the folder.
- Not enough data in file to identify what happened.
- There is nothing in file regarding misuse.

How SSA Discovered the Misuse

Table D-15 presents information about how SSA discovered the misuse: such information was available for 180 of the 291 cases. For about 10

TABLE D-15 Source of Misuse Allegations

Source	Number	Percent
Beneficiary	30	10.3
Third party	75	25.8
New payee	31	10.7
Combination of above	4	1.4
Other	40	13.8
Not available	111	38.1

SOURCE: Data generated from a review of SSA administrative data of identified misusers conducted for the National Academies Committee on Social Security Representative Payees (2006).

percent of the cases in our in-depth study, the misuse was detected through allegations from the beneficiary. For another 36 percent, the allegation came from a third party, such as a family member, a friend of the beneficiary, a neighbor, or a doctor. In third-party situations, the scenario is often that the custodial payee comes to the local SSA field office and complains about nonreceipt of money for the care of a child from the noncustodial parent. The noncustodial parent will falsely have stated to SSA that he or she has physical custody of the beneficiary. As shown in the table, the information sometimes comes to light when there is a change in payee or from a combination of the beneficiary and others.

Summary

We undertook the in-depth study of randomly selected identified misusers to look for information about the demographic and socioeconomic characteristics of the payees and their beneficiaries or other factors that might explain misuse. We were also looking for circumstances that led to the misuse event.

We note first that the administrative data do not provide a complete picture of misuse for a variety of reasons. First, folders are not available for all beneficiaries. Some folders have been destroyed because the files are inactive. Others are kept in either the field or the payment centers and cannot be readily retrieved. Second, for those folders we did retrieve and review, many records were incomplete and had limited or no data on misuse. The rest of this section summarizes what we learned from the available information.

Circumstances of Misuse

- About 50 percent of the documented misuse happens because the payee does not have the beneficiary in his or her care (custody) as claimed.
- Most custodial misuse involves parents.
- About 5 percent of the documented misuse involves money in dedicated accounts.
- More than 12 percent of the documented misuse results from payees' not having reported to SSA events affecting the benefit amount or entitlement to the benefit.

How SSA Learns about Misuse

- SSA is most likely to learn about misuse through a third party.
- The beneficiaries also tell SSA about misuse.

Amount Misused

- The average misuse amount was $4,414.
- The smallest amount misused was $41 (benefits for 1 month).
- The highest single amount misused was $45,000 over 9 years and 4 months.

Length of Misuse

- Most misuse involves just 1 month of benefits.
- More than 77 percent of misuse lasts no more than 12 months.

Recovery of Misused Funds

- The misused funds are recovered (or in the process of being recovered) in only a small number of cases.

Characteristics of Payees

- Most are female.
- About 35 percent are themselves beneficiaries; they are most likely to be receiving disability benefits.
- About 12 percent have more than one address in a year.
- More than 80 percent provide a phone number at the time of the application.
- Close to 72 percent give the same mailing and residence address.
- They speak the same language as their beneficiaries.

- About 5 percent have criminal records.
- About 7 percent are reported to be abusing drugs or alcohol.

Characteristics of Beneficiaries

- The majority of the misused beneficiaries are male.
- The majority of the misused beneficiaries are children (55 percent).
- The beneficiary is frequently the son or daughter of the payee.
- About 32 percent receive OASDI disability benefits.
- Close to 35 percent are SSI beneficiaries.
- Almost all beneficiaries speak the same language as their payees.
- Only a small percentage has alleged drug or alcohol abuse (4 percent).
- About 25 percent suffer from alleged mental illness.
- Less than 2 percent have been judged legally incompetent.

Characteristics Associated with Payeeship

- About 9 percent of payees have failed to respond to the annual accounting process.
- About 24 percent of the payees serve as payees for several beneficiaries at the time of the misuse.
- In spite of being labeled misusers in the RPS, about 9 percent are still serving as payees.
- About 44 percent of the misuser payees have been terminated once, 16 percent twice, and 22 percent more than twice with the notation that "a more suitable payee" has been found for either the beneficiaries associated with the misuse or for other beneficiaries served by the payees.

CONCLUSION

Misuse is not very well documented in SSA data. When SSA staff makes a notation of misuse in the RPS, they do not necessarily document the circumstances that led to the misuse determination. When it is documented, it appears that other SSA staff would on several occasions have disagreed with the decision and instead labeled the misuse as an overpayment.

Without documentation, it is difficult for SSA to learn about either the demographic and socioeconomic characteristics of the payees who get labeled as misusers in the RPS or the causes of misuse. As a consequence, it is difficult for SSA to make empirically based decisions about changes to the policies and procedures pertaining to selection, training, and monitoring representative payees.

Reference

U.S. Social Security Administration
    2002   *Representative Payee Program.* Office of Income Security Programs. Baltimore, MD: U.S. Social Security Administration.

Appendix E

Current Annual Accounting Form

The current two-page accounting form used in the Representative Payee Program is reproduced in this appendix.

```
6232
```

Representative Payee Report

Social Security Administration, P.O. Box 6232, Wilkes-Barre, PA 18767-6232

FORM APPROVED
OMB NO. 0960-0068

PAYEE'S NAME AND ADDRESS

REPORT PERIOD

SOCIAL SECURITY NUMBER

FROM   TO

BENEFICIARY   FP

ID   BIC   D   TP   CX   GS   P2   DOC

CF   TAA   FF   BSSN

If change of address, correct and check box. ☐

This report is about the benefits you received between   and   for the beneficiary,   Please read the enclosed instructions before completing this form to help you answer each question.

1. Were you (the payee) convicted of a crime considered to be a felony between   and   ?
If YES, please explain in REMARKS on the back of this form.

YES ☐   NO ☐

2. Did the beneficiary continue to live alone, or with the same person, or in the same institution from   to   ? If NO, please explain and provide the beneficiary's current address in REMARKS on the back of this form.

☐   ☐

3. Benefits paid to you between   and   = $
Benefits you reported as **saved** on last year's report. = $

Total Accountable Amount = $

YES ☐   NO ☐

A. Did you (the payee) decide how the   was spent or saved? →
If NO, please explain in REMARKS on the back of this form.

☐   ☐

DOLLAR AMOUNT
(NO CENTS)

B. How much of the   did you spend for the beneficiary's food and housing between   and   ? →

☐☐☐ , ☐☐☐

C. How much of   did you spend on other things for the beneficiary such as clothing, education, medical and dental expenses, recreation, or personal items between   and   ? →

☐☐☐ , ☐☐☐

D. How much, if any, of the   did you save for the beneficiary as of   ? If none, show zeros. →

☐☐☐ , ☐☐☐

4. If you showed an amount in 3.D. above, place an "X" in the boxes below to show how you are saving the benefits. If you have more than one account, you may mark more than one box in each section.

A. TYPE OF ACCOUNT					B. TITLE OF ACCOUNT		
Savings/ Checking Account	U.S. Savings Bonds	Certificates of Deposit	Collective Savings/ Checking Account	Other	Beneficiary's Name by Your Name	Your Name for Beneficiary's Name	Other
☐	☐	☐	☐	☐	☐	☐	☐

FORM SSA-623-OCR-SM (12-2004)

Continued on the Reverse →

```
┌─────────┐
│ 6232B   │   ▐▊▌▐▊▌▐▊▌
└─────────┘
```

5. A. If you answered **"OTHER"** in 4.A. on the front page, show the type of account or investment in which the benefits are saved. ⟶

TYPE OF ACCOUNT

B. If you answered **"OTHER"** in 4.B. on the front page, show the title of the account in which the benefits are saved. ⟶

TITLE OF ACCOUNT

REMARKS

I declare under penalty of perjury that I have examined all the information on this form, and on any accompanying statements or forms, and it is true and correct to the best of my knowledge. I understand that anyone who knowingly gives a false or misleading statement about a material fact in this information, or causes someone else to do so, commits a crime and may be sent to prison, or may face other penalties, or both.

PAYEE'S SIGNATURE
(If signed by mark (X), two witnesses must sign below)

6.

DATE

8.

PRINT RELATIONSHIP TO BENEFICIARY OR TITLE

7.

DAYTIME TELEPHONE NUMBER(S)
(Include area code)

9. Area Code

WITNESS SIGNATURES ARE REQUIRED ONLY IF THE PAYEE'S SIGNATURE ABOVE HAS BEEN SIGNED BY MARK (X).

SIGNATURE OF WITNESS

DATE

SIGNATURE OF WITNESS

DATE

FORM **SSA-623-OCR-SM** (12-2004)

Appendix F

Proposed Representative
Payee Annual Report

As discussed in Chapter 4, the committee proposes a revision to the Social Security Administration's accounting form. This appendix presents that proposed two-page annual report.

Representative Payee Report

FORM APPROVED

OMB NO. XXXX-XXXX

Social Security Administration, P.O. Box 6231, Wilkes-Barre, PA 18767-9912

This report is about the benefits you received for the beneficiary, BENEFICIARY NAME, between MM/DD/YYYY and MM/DD/YYYY. Please use the enclosed instructions to help answer each question.

Representative Payee Name and Address Label info	Report Period - not clear how this helps respondent here FROM: MM/DD/YYYY TO: MM/DD/YYYY

BAR CODE WITH SSN OF BENEFICIARY?

Has your address changed? (put Q in bold, and either add a number or an arrow) .
[ ] Yes ---→ Write your new address here: (put the address lines in) _____
[ ] No   _____
                    _____

---

1. **For a typical or average month between MM/DD/YYYY and MM/DD/YYYY:**   WHOLE DOLLAR AMOUNT
                                                  (NO CENTS)

  A. How much *per month* of the benefits received were spent on the beneficiary's **food**?   □,□□□
How much of the benefit money do you spend for the beneficiary's **food** each month, on average?

  B. How much *per month* of the benefits received were spent on the beneficiary's **housing**?   □,□□□
Same suggested rewording

  C. How much *per month* of the benefits received were spent on other things for the beneficiary   □,□□□
    such as **clothing, education, medical and dental expenses, recreation, or personal items**?
Same suggested rewording

2. Are the monthly Social Security benefits for BENEFICIARY NAME received by direct deposit?   YES NO
                                                                            □   □

3. Do you have any accounts where you save **Social Security** benefits for BENEFICIARY NAME?   YES NO
    If YES, please answer questions 4 and 5. If NO, go to Question 6 on back of form.   □   □

4. What was the **TOTAL** balance of Social Security savings as of MM/DD/YYYY for   WHOLE DOLLARS
   BENEFICIARY NAME?   □□,□□□

5. Please mark yes or no to indicate the types of accounts used for savings from BENEFICIARY NAME Social Security benefits and provide information about the name on each account (title of account) and the account balance.
                                          Yes   No   Title of Account   Current Balance (whole
dollars)
      SAVINGS/CHECKING ACCOUNT?   [ ]   [ ]
      US SAVINGS BONDS ?
      CERTIFICATE OF DEPOSIT?
      STOCKS/BONDS/MUTUAL FUNDS ?
      OTHER -Specify _____

Question 5 REMARKS or OTHER SAVINGS ACCOUNTS with TITLE OF ACCOUNT:

6. Is the address where you (the payee) get your mail different from your residence?　　　YES　NO
　　□　　□

7. How many times have you changed residence in the past 2 years?　　　　　　　　　_____

8. What were your (the payee) primary **HOUSEHOLD** sources of income between MM/DD/YYYY and MM/DD/YYYY?　　　　(You may check as many as appropriate.)
Put into a single column so all options are seen
　　□ **EMPLOYED BY SELF**　　　□ **PENSION OR VETERAN'S BENEFITS**
　　□ **EMPLOYED BY OTHER**　　□ **TEMPORARY ASSISTANCE FOR NEEDY FAMILIES**
　　□ **SSA BENEFITS**　　　　　□ **OTHER SOURCES (SPECIFY)** _____

9. Currently, how many beneficiaries do you manage Social Security benefits for?　　_____

10. In the past 2 years, for how many beneficiaries have you **STOPPED** managing their benefits?　　_____

11. What is your relationship to BENEFICIARY NAME (Check only 1)
　　Single column or at least move 'none of these' to end of second column
　　□ **PARENT/ADOPTIVE PARENT**　　　□ **SPOUSE**
　　□ **CHILD/ADOPTIVE CHILD**　　　　□ **GRANDPARENT**
　　□ **OTHER RELATIVE**　　　　　　　□ **NON RELATIVE FRIEND**
　　□ **NONE OF THESE**

12. Are you a court appointed guardian or custodian?　　　　　　　　　　　　　YES　NO
　　　　　　　　　　　　　　　　　　　　　　　　　　　　　　　　　　　　□　　□

13. Have you been convicted of a felony in the past 2 years?　　　　　　　　　YES　NO
　　　　　　　　　　　　　　　　　　　　　　　　　　　　　　　　　　　　□　　□

14. Does BENEFICIARY NAME currently live with you?　　　　　　　　　　　YES　NO
　　　If NO, please explain and provide the beneficiary's current address in REMARKS　□　　□
　　　below.

REMARKS
Please refer to the item number when making your remarks.

I CERTIFY THAT THE INFORMATION I HAVE GIVEN ON THIS FORM IS TRUE. (A PERSON WHO CONCEALS OR FAILS TO TELL SSA ABOUT "EVENTS" ASKED ABOUT ON THIS FORM WITH THE INTENT TO FRAUDULENTLY RECEIVE BENEFITS MAY BE FINED, IMPRISONED, OR BOTH.)

PAYEE'S SIGNATURE	**DATE**
PAYEE DAYTIME TELEPHONE NUMBER(s) (*Include area code*)	
PAYEE EMAIL ADDRESS (we will not release this address to others)	
WITNESS SIGNATURES ARE REQUIRED ONLY IF THE PAYEE'S SIGNATURE ABOVE HAS BEEN SIGNED BY MARK (X)	
SIGNATURE OF WITNESS 1	**DATE**
SIGNATURE OF WITNESS 2	**DATE**

Appendix G

Biographical Sketches of Committee Members and Staff

BARBARA A. BAILAR (*Chair*) is recently retired from the National Opinion Research Center (NORC) and now consults on survey methodology. Immediately prior to joining NORC, she was the executive director of the American Statistical Association. Most of her career was spent at the U.S. Census Bureau where she was the associate director for Statistical Standards and Methodology. She has published numerous articles in such journals as *JASA, Demography*, and *Survey Research Methods*. She is a past president of the American Statistical Association and the International Association of Survey Statisticians, as well as a past vice president of the International Statistical Association. She is an elected fellow of the American Statistical Association and the American Association for the Advancement of Science. She received a Ph.D. in statistics from American University in Washington, D.C.

NANCY COLEMAN is a consultant in philanthropy, aging and policy, and she served as a program officer at the Harry and Jeanette Weinberg Foundation, Inc., in 2005. Previously, she was director of the American Bar Association's Commission on Law and Aging. She was an investigator for the U.S. Senate Special Committee on Aging and project director of Citizens for Better Care. She has published extensively on issues facing the elderly, including legal and financial concerns, aging, spirituality, and religion. She served as a U.S. delegate to The Hague Conference on Private International Law and helped draft an international agreement on the recognition of incapacitated adults. She also served as chair of the Social

Security Administration's Representative Payment Advisory Committee. She received an M.S.W. and an M.S. in political science from the University of Michigan.

CATHRYN S. DIPPO retired as associate commissioner of the Office of Survey Methods Research at the Bureau of Labor Statistics (BLS) of the U.S. Department of Labor. While at BLS, she chaired the FedStats R&D Working Group and the Current Population Survey Redesign. She started the National Science Foundation/American Statistical Association/BLS Senior Research Fellow Program in the mid-1980s and the BLS Behavioral Science Research Center in the late 1980s. An office holder and member of several statistical societies, she has published a number of articles and has served as a referee for various statistical journals. She received a Ph.D. in mathematical statistics from George Washington University.

CARROLL L. ESTES is professor of sociology at the University of California, San Francisco (UCSF). She is the founding and former director of the Institute for Health and Aging and the former chair of the Department of Social and Behavioral Sciences of the School of Nursing at UCSF. She is a member of the Institute of Medicine. She has served as a consultant to the U.S. Commissioner of Social Security and to U.S. Senate and House committees for more than two decades. She investigates the effects of fiscal austerity and social policy on the elderly and the agencies and institutions that serve them, and her research has been published in numerous journals. She received a Ph.D. in sociology from the University of California at San Diego.

TIMOTHY P. JOHNSON is director of the Survey Research Laboratory, professor of public administration, and research professor of epidemiology and biostatistics at the University of Illinois at Chicago. He teaches courses in sample design, research methodology, and multivariate statistical analysis, and he is currently serving as cochair of the university's Social and Behavioral Sciences Institutional Review Board. His recent work has focused on the social epidemiology of substance use and measurement errors in survey research, with an emphasis on the effects of respondent culture. He has published approximately 80 peer-reviewed papers, and he is a member of the editorial board of the journal *Substance Use and Misuse*. He received a Ph.D. in sociology from the University of Kentucky.

JEFFREY LUBBERS is a fellow in law and government at American University's Washington College of Law, where he teaches administrative law and related courses. He has served in various positions with the Administrative Conference of the United States (ACUS), the U.S. government's advisory agency on procedural improvements in federal programs. He was ACUS' research director and developed ideas for new studies, and he assisted committees in developing recommendations from the studies on a wide variety of administrative law subjects, including the Representative Payee Program. He has also worked with congressional committees and agencies to seek implementation of ACUS recommendations and served as team leader for Vice President Gore's National Performance Review Team on Improving Regulatory Systems. He is a member of the bars of Maryland and the District of Columbia. He received a J.D. from the University of Chicago Law School.

SARAH NUSSER is a professor in the Department of Statistics and affiliated with the Center for Survey Statistics and Methodology at Iowa State University and she previously served as director of the Center. She is a fellow of the American Statistical Association (ASA), and she has served as chair of the ASA's Survey Research Methods Section and as a member of ASA advisory committees, including the Survey Review Committee and the Behavioral Risk Factor Surveillance System Advisory Group. Her research interests include computer-assisted survey methods, sample design and estimation for natural resource and social surveys, accuracy assessment of spatial databases, and social policy applications including welfare reform evaluation and estimation of dietary intake distributions. As a faculty member in the Center, she consults with a wide range of researchers on survey statistics and methodologies for conducting surveys. She received a Ph.D. in statistics from Iowa State University.

ROBERT SANTOS is senior institute methodologist at The Urban Institute in Washington, DC. His prior positions include executive vice president of NuStats, vice president of statistics and methodology at NORC at the University of Chicago, and director of survey operations at the Survey Research Center at the University of Michigan at Ann Arbor. His professional credits include more than 40 reports and papers and leadership roles in survey research associations. He has served as a member of the Census Advisory Committee of Professional Associations and the editorial board of the *Public Opinion Quarterly,* and he has held numerous elected and appointed leadership positions in both the ASA and the American Association for Public Opinion Research. He is a fellow of the ASA and a recipient of the 2006 ASA Founder's Award for excellence in survey statistics and contributions to the statistical community. He received an M.A. in statistics from the University of Michigan.

PAMELA B. TEASTER is an associate professor at the Graduate Center for Gerontology and Department of Health Behavior in the College of Public Health at the University of Kentucky, Lexington. She serves as a commissioner on the Commission on Law and Aging of the American Bar Association and is vice president of the National Committee for the Prevention of Elder Abuse. She is a former editor of the *Journal of Elder Abuse and Neglect* and serves on the editorial board of the *Journal of Applied Gerontology*. She is an expert in adult protective services, guardianship, victimization of older women, and sexual abuse, and she has published numerous articles in scholarly journals. She received a Ph.D. in public administration and public affairs from Virginia Polytechnic Institute.

STAFF

CHARLES (BUD) PAUTLER *(study director)* has been on staff since January 2005. He was formerly the director of research of the Small Business/Self-Employed Division of the Internal Revenue Service, and previously served in three other agencies, including the U.S. Census Bureau. He earned a Ph.D. in mathematical statistics from the George Washington University.

KIRSTEN WEST *(senior program officer)* is working at the National Academies on an Intergovernmental Personnel Act Agreement with the U.S. Census Bureau. Her area of expertise is census coverage error measurement. She earned a Ph.D. in sociology from the University of North Carolina at Chapel Hill.

LINDA DePUGH *(administrative assistant)* has worked in a variety of capacities over a long career for many committees at the National Research Council and Institute of Medicine including the Strategic Planning Advisory Group for Education, the Committee on Scientific Principles for Education Research, the Committee on the Impact of the Changing Economy on the Education System, the Committee on Monitoring International Labor Standards, and the Committee on Analyzing the U.S. Content of Imports and Foreign Content of Exports. She has an A.A. degree in business from the Durham Business School in North Carolina.